GEOPHYSICAL MONOGRAPH SERIES

David V. Fitterman, managing editor
Laurence R. Lines and Patrick Daley, volume editors

NUMBER 14

EDGE AND TIP DIFFRACTIONS

THEORY AND APPLICATIONS
IN SEISMIC PROSPECTING

Kamill Klem-Musatov
Arkady M. Aizenberg
Jan Pajchel
Hans B. Helle

Society of Exploration Geophysicists
The international society of applied geophysics
Tulsa, Oklahoma, U.S.A.

ISBN 978-0-931830-56-3 (Series)
ISBN 978-1-56080-149-8 (Volume)

Society of Exploration Geophysicists
P. O. Box 702740
Tulsa, OK 74170-2740

Published 2008
Printed in the United States of America

Library of Congress Cataloging-in-Publication Data

Edge and tip diffractions : theory and applications in seismic prospecting / Kamill Klem-
Musatov ... [et al.].
     p. cm. -- (Geophysical monograph series ; no. 14)
  Includes bibliographical references and index.
  ISBN 978-1-56080-149-8 (volume) -- ISBN 978-0-931830-56-3 (series)
  1. Geometrical diffraction. 2. Seismic prospecting--Mathematical models. 3. Waves--
Diffraction. 4. Seismic waves. I. Klem-Musatov, K. D. (Kamill Davydovich)
  TN269.83.E34 2008
  622'.1592015118--dc22
                              2008054538

Contents

About the Authors

Kamill Klem-Musatov is one of the principal researchers at the Trofimuk Institute of Petroleum Geology and Geophysics of the Siberian Branch of the Russian Academy of Sciences in Novosibirsk and is professor of geophysics at Novosibirsk State University. In the 1970s, he obtained asymptotic formulas for waves diffracted by edges and vertices on seismic interfaces, originally presented in Russian and later in English. Klem-Musatov's approach found acceptance in the international geophysical community and has been applied to solve various seismic problems.

Klem-Musatov graduated from the Moscow Mining Institute in 1958, received a Ph.D. in the theory of elasticity in 1967, and earned a doctor of science in geophysics in 1981.

Arkady M. Aizenberg is senior scientific researcher at the Trofimuk Institute of Petroleum Geology and Geophysics of the Siberian Branch of the Russian Academy of Sciences in Novosibirsk. Since the early 1980s, he has been at the forefront of the development of an analytic description of seismic wavefields in 3D heterogeneous layered and block media. He introduced a concept of a tip-wave beam composed from reflected/transmitted waves and associated edge and tip waves. In the early 2000s, Aizenberg introduced the generalization of the Weyl spectral representation for wavefields propagating near curved interfaces in heterogeneous acoustic and elastic media.

Aizenberg graduated from Novosibirsk State University in 1971 and received a Ph.D. in geophysics in 1994 from the Supreme Certifying Commission of the Russian Federation.

Jan Pajchel is a principal geophysicist in Statoil-Hydro Research Centre, Bergen, Norway, where he is working on 3D seismic modeling and imaging. After six years as a postdoctoral fellow/lecturer at the Seismological Observatory, University of Bergen, he joined the Research Centre of Norsk Hydro in 1987. From 1993 through 2001, Pajchel was the industrial partner of four and coordinator of two European Commission Projects in the framework of the Joint Opportunities for Unconventional or Long-term Energy Supply (JOULE) program on nonnuclear energy sources.

Pajchel received an M.Sc. in solid-earth physics from the University of Warsaw in 1969. For the next 14 years, his time was devoted to crustal studies in the Deep Seismic Soundings Laboratory of the Polish Academy of Sciences, where he received a Ph.D. in 1976. He is a member of SEG.

Hans B. Helle is an adviser in petroleum geophysics at Odin Petroleum AS, a consultancy firm in Bergen, Norway, which he joined in 2006. From 1970 through 1978, he was a research assistant and associate scientist at the Geophysical Institute of the University of Bergen. From 1978 through 1980, he was a research engineer at SINTEF in Trondheim, Norway. Helle was an associate at the Seismological Observatory at the University of Bergen from 1980 through 1983 and then served as a research geophysicist at the Norsk  Hydro E & P Research Centre in Bergen for 22 years. He is the author or coauthor of numerous papers on oceanography, seismology, and geophysics and has served as an associate editor of GEOPHYSICS and *Geophysical Prospecting*. His main current interests are numerical modeling of seismic wave propagation and its application to rock and fluid characterization from seismic data.

Helle received an M.Sc. in oceanography from the University of Bergen in 1973. He is a member of AGU, EAGE, SEG, and SPE.

Preface

This book consists in part of lecture notes based on joint research results of the Russian Academy of Sciences Institute of Geophysics in Novosibirsk and the Norsk Hydro Research Centre in Bergen, Norway. The text focuses on edge and vertex diffraction phenomena, which are of significant importance in seismic prospecting. The material presented here is collected from several previous publications but also includes new material taken from unpublished technical reports.

Modeling of diffracted waves can be used to solve practical problems, depending on the character of the problem. In some cases, the minute consideration of an individual diffracted wave can be important. Such detailed description involving heavy mathematical formalism is considered in Klem-Musatov (1994). However, of greatest practical importance is the case when diffracted waves can be regarded as arrivals complicating regular reflections/transmissions that represent basic geophysical information.

In such situations, it is possible to use the simplest description of diffraction phenomena in the shadow-boundary neighborhoods of regular reflections/transmissions (the boundary-layer approximation). Because of its simplicity and practical implication, we deal only with such an approximation here. We explain the concepts of wave propagation and diffraction theory in simplified terms. All concepts are introduced for the scalar wave-equation case to reduce mathematical constructions. Final formulas are generalized to the vector description using decomposition of vectors over the scalar basis.

The introduction deals only with the concepts of asymptotic ray theory (ART) that are necessary for the following chapters. There are two specific reasons for presenting these concepts. First, ART follows directly from the equation of motion and from its asymptotic property (Courant and Hilbert, 1966). Therefore, there is no need to deduce ART from the ray series. Second, the reflection/transmission problem is stated in a nontraditional form suitable for introducing the general concept of the edge-diffracted wave. We refer to Červený (2001) for a comprehensive treatment of the ray-series approach.

The main concepts of edge diffraction (the shadow-boundary neighborhood and the boundary-layer approximation) are introduced in Chapter 2.

Using these concepts, we show that the ART wave and the edge-diffracted wave correspond to a pair of linearly independent asymptotic solutions of the wave equation.

Modification of ART to include edge waves is introduced in Chapter 3. The edge wave is considered as a smoothing correction for a discontinuity in the ART wavefield at the shadow boundary. To obtain this correction, we use the analytic continuation technique of the singular integral equation. Such an approach leads to the general description of edge waves of any kind (acoustic, elastic, electromagnetic, and so forth) without considering a particular form of the equation of motion.

The concept of vertex diffraction is introduced in Chapter 4 as a correction to the discontinuities of edge waves at secondary shadow boundaries caused by vertices. Here again, the analytic continuation technique is used to find smoothing corrections. This also leads to the general formulas for vertex waves of any kind.

The simplicity of edge-wave theory makes it easy to use for modeling in seismic prospecting (Landa et al., 1987; Frøyland et al., 1988; Pajchel et al., 1988; Hoffmann et al., 1993; Klaeschen et al., 1994; Hron and Chan, 1995; Wang and Waltham, 1995; Luneva, 1996; Gallop and Hron, 1998; Landa and Keydar, 1998). The importance of enhancing diffractions to improve the resolution in conventional seismic processing is the topic of a more recent paper (Khaidukov et al., 2004, p. 1490): "Diffractions have been the stepchildren of traditional processing/imaging. However, if they receive the attention they deserve, we will be able to see the invisible."

The edge-wave approach also is used for diffraction modeling in a 2D modeling software package developed for Norsk Hydro and the University of Bergen (Pajchel et al., 1987). In Chapter 5, we demonstrate its application in simple 2D models and in more realistic geologic situations involving complex tectonics. Synthetic data generated by this method are suited well for tests of various procedures in seismic processing. In particular, the presence of edge-diffracted waves is a minimum requirement in the testing of seismic-migration procedures. Results of prestack and poststack migration tests on synthetic data indicate that diffraction computations by edge-wave theory are sufficiently accurate for applications in practical exploration problems.

Edge-wave and tip-wave formulas fail where the ray field changes rapidly, e.g., in caustic zones. The last section of the book deals with the edge/tip-wave superposition method, which considers caustic phenomena. The above shortcomings are eliminated in this form. We briefly outline the background theory of the tip-wave single-superposition method for computation of seismic waves in 3D media and its implementation on a computer. Then we present numerical results from experiments on realistic models with faults and synclines.

The notes used in this book are based also on master's-level courses taught by Kamill D. Klem-Musatov at Novosibirsk State University. Since the first edition of these lecture notes was printed in 1994, comments and corrections from students have been incorporated into the present edition.

—Kamill D. Klem-Musatov
Arkady M. Aizenberg
Jan Pajchel
Hans B. Helle
Novosibirsk and Bergen, November 2008

Acknowledgments

This work has been supported by Norsk Hydro a.s. through the 3D Seismic Modeling project and by the Norwegian Research Council under the NATO Senior Scientist Visiting Program (K. D. Klem-Musatov). The Russian Fund for Basic Research (grants 03-05-64941 and 07-05-00671) and the Federal Agency for Science and Innovations of the Russian Federation (grant RI-112/001/252) have provided significant background research support. We are grateful to Brian Farrelly for his suggestions for improving the manuscript before submission and to Odin Petroleum AS for financial support in the final editing stage.

We also appreciate the people who assisted with editing and production of the book for SEG: David V. Fitterman, managing editor of the SEG Geophysical Monograph Series; Laurence R. Lines and Patrick Daley, volume editors; Jennifer Baltz, copy editor; and Ted Bakamjian, Jennifer Cobb, and Rowena Mills of the SEG publications department.

Chapter 1

Introduction

Basic concepts of wave-propagation theory

Propagating waves represent an aspect of more general forms of motion that can be described by the exact equation of continuum mechanics. Although a study of general forms of motion is practically possible only by means of numerical methods, their wave aspects can be expressed in terms of an elementary theory. The basic concepts of such a theory were established as a generalization of experimental facts long before mechanics itself appeared as a branch of mathematical physics. Only much later were they derived as consequences of the exact equations of mechanics. This system of concepts has an important property — it allows us to build a simple wave-propagation theory common to waves of any kind. That is why the basic concepts of wave-propagation theory can be introduced by considering the simplest equation of motion and then generalizing to more complex situations.

Here we consider these concepts in the case of the scalar wave equation

$$\sum_{i=1}^{3} \frac{\partial^2 f^*}{\partial x_i^2} - \frac{1}{c^2} \frac{\partial^2 f^*}{\partial t^2} = F^*, \tag{1}$$

where f^* and F^* are functions of space coordinates (x_1, x_2, x_3) and time t and c is a function of space coordinates only.

Canonical coordinate transformation

The basic concepts of wave-propagation theory arise in investigating the wave equation written in special coordinates (e.g., Courant and Hilbert, 1966). Here we consider the corresponding coordinate transformation.

1

Introduce new variables in wave equation 1:

$$y_i = y_i(x_1, x_2, x_3, x_4) \quad \text{with} \quad i = 1, 2, 3, 4, \tag{2}$$

where we temporarily denote $t = x_4$ for brevity. To do this, consider a solution of the wave equation as a composite function

$$f^* = f^*[y_1(x_1, \ldots x_4), \ldots y_4(x_1, \ldots x_4)].$$

Differentiate this expression as a composite function

$$\frac{\partial f^*}{\partial x_i} = \sum_{k=1}^{4} \frac{\partial f^*}{\partial y_k} \frac{\partial y_k}{\partial x_i},$$

$$\frac{\partial^2 f^*}{\partial x_i^2} = \sum_{k=1}^{4} \left(\sum_{m=1}^{4} \frac{\partial^2 f^*}{\partial y_k \partial y_m} \frac{\partial y_k}{\partial x_i} \frac{\partial y_m}{\partial x_i} + \frac{\partial f^*}{\partial y_k} \frac{\partial^2 y_k}{\partial x_i^2} \right). \tag{3}$$

Solving equation 2 for the old variables and substituting the results into equations 3, express the derivatives under consideration using the new variables.

To do this, choose the coordinate transformation as follows:

$$y_i = x_i \quad \text{with} \quad i = 1, 2, 3, \tag{4}$$

$$y_4 = x_4 - \tau(x_1, x_2, x_3) = t - \tau(y_1, y_2, y_3), \tag{5}$$

where τ is an arbitrary function and variable x_4 again is denoted by t. Then the following relationships hold true with $k = 1, 2, 3$,

$$\frac{\partial y_k}{\partial x_i} = 1 \quad \text{for} \quad k = i, \qquad \frac{\partial y_k}{\partial x_i} = 0 \quad \text{for} \quad k \neq i,$$

$$\frac{\partial^2 y_k}{\partial x_i^2} = 0 \quad \text{for} \quad i = 1, 2, 3, 4,$$

and for $k = 4$,

$$\frac{\partial y_4}{\partial x_i} = \frac{\partial y_4}{\partial y_i} = -\frac{\partial \tau}{\partial y_i}, \quad \frac{\partial^2 y_4}{\partial x_i^2} = -\frac{\partial^2 \tau}{\partial y_i^2} \quad \text{for} \quad i = 1, 2, 3,$$

$$\frac{\partial y_4}{\partial x_4} = 1, \quad \frac{\partial^2 y_4}{\partial x_4^2} = 0.$$

Substitute these expressions into equations 3:

$$\frac{\partial^2 f^*}{\partial x_i^2} = \frac{\partial^2 f^*}{\partial y_i^2} + \left(\frac{\partial \tau}{\partial y_i}\right)^2 \frac{\partial^2 f^*}{\partial y_4^2} - 2\frac{\partial \tau}{\partial y_i}\frac{\partial^2 f^*}{\partial y_i \partial y_4} - \frac{\partial^2 \tau}{\partial y_i^2}\frac{\partial f^*}{\partial y_4} \quad \text{for} \quad i = 1, 2, 3,$$

$$\frac{\partial^2 f^*}{\partial x_4^2} = \frac{\partial^2 f^*}{\partial t^2} = \frac{\partial^2 f^*}{\partial y_4^2}.$$

Introduce these expressions into equation 1, considering the relationships

$$\sum_{i=1}^{3}\frac{\partial^2 f^*}{\partial y_i^2} = \Delta f^*, \quad \sum_{i=1}^{3}\frac{\partial \tau}{\partial y_i}\frac{\partial}{\partial y_i}\left(\frac{\partial f^*}{\partial y_4}\right) = \nabla\tau\cdot\nabla\left(\frac{\partial f^*}{\partial y_4}\right),$$

$$\sum_{i=1}^{3}\left(\frac{\partial \tau}{\partial y_i}\right)^2 = (\nabla\tau)^2, \quad \sum_{i=1}^{3}\frac{\partial^2 \tau}{\partial y_i^2} = \Delta\tau,$$

and write the wave equation as

$$\Delta f^* + \left[(\nabla\tau)^2 - \frac{1}{c^2}\right]\frac{\partial^2 f^*}{\partial y_4^2} - 2\nabla\tau\cdot\nabla\left(\frac{\partial f^*}{\partial y_4}\right) - \Delta\tau\frac{\partial f^*}{\partial y_4} = F^*. \quad (6)$$

Now choose function τ as a solution of the following eikonal equation

$$(\nabla\tau)^2 = \frac{1}{c^2}. \quad (7)$$

Then the coefficient of the second derivative in equation 6 becomes zero, and the wave equation can be written as

$$\Delta f^* - 2\nabla\tau\cdot\nabla\left(\frac{\partial f^*}{\partial y_4}\right) - \Delta\tau\frac{\partial f^*}{\partial y_4} = F^*. \quad (8)$$

A coordinate transformation making the coefficient of the second derivative zero is called canonical.

Nonstationary wave

Variable 5, with the additional condition 7, considered as a surface in the 4D space (x_1, x_2, x_3, t), is called a characteristic of the wave equation.

The important property of characteristics is stated by the following theorem:

> Let functions $f^*(x_1, x_2, x_3, t)$ be solutions of equation 1, which are continuous with all their derivatives on some surface $S(x_1, x_2, x_3, t)$. If a solution exists such that on S, its second derivative along the normal to S has a discontinuity of the first order with all other derivatives being continuous, then the surface S is a characteristic of equation 1.

To prove this theorem, transform equation 1 to the form 6, introducing the new variables given by equations 4 and 5. The solution of such an equation satisfies the conditions of the theorem at any point on the surface $y_4 =$ constant only if condition 7 is satisfied. This means that the surface $y_4 =$ constant is a characteristic of equation 1. The theorem is proven.

Consider a possible type of solution satisfying the conditions of the theorem and leading to the basic concepts of wave-propagation theory. Take

$$f^*(y_1, y_2, y_3, y_4) \equiv 0 \quad \text{when} \quad y_4 < 0,$$

$$f^*(y_1, y_2, y_3, y_4) \neq 0 \quad \text{when} \quad y_4 \geq 0. \tag{9}$$

Then

$$\frac{\partial f^*}{\partial y_i} = 0, \quad \frac{\partial^2 f^*}{\partial y_i \partial y_k} = 0, \quad F^* = 0 \quad \text{when} \quad y_4 < 0 \quad \text{for} \quad i, k = 1, 2, 3, 4.$$

This type of solution satisfies the conditions of the theorem if the following limits apply:

$$\lim_{\varepsilon \to 0} \left(f^* \right)_{y_4=\varepsilon} = 0, \quad \lim_{\varepsilon \to 0} \left(\partial f^*/\partial y_i \right)_{y_4=\varepsilon} = 0 \quad \text{for} \quad i = 1, 2, 3, 4,$$

$$\lim_{\varepsilon \to 0} \left(\partial^2 f^*/\partial y_i^2 \right)_{y_4=\varepsilon} = 0, \quad \lim_{\varepsilon \to 0} \left(\partial^2 f^*/\partial y_i \partial y_4 \right)_{y_4=\varepsilon} = 0 \quad \text{for} \quad i = 1, 2, 3, \tag{10}$$

$$\lim_{\varepsilon \to 0} \left(\partial^2 f^*/\partial y_4^2 \right)_{y_4=\varepsilon} = \varphi, \tag{11}$$

where ε is a small positive quantity and φ is the discontinuity of the solution's second derivative along the normal to the characteristic $y_4 = 0$. Let the following derivatives be continuous as well:

$$\lim_{\varepsilon \to 0} \left(\partial^3 f^* / \partial y_4 \partial y_i^2 \right)_{y_4 = \varepsilon} = 0 \quad \text{for} \quad i = 1, 2, 3, \quad \lim_{\varepsilon \to 0} \left(\partial F^* / \partial y_4 \right)_{y_4 = \varepsilon} = 0. \tag{12}$$

Substituting equation 5 into 9, write the solution under consideration in the form of the nonstationary wave

$$f^* = \begin{cases} 0 & \text{when} \quad t < \tau, \\ f^*(y_1, y_2, y_3, t - \tau) & \text{when} \quad t \geq \tau, \end{cases} \tag{13}$$

where the function $\tau = \tau(y_1, y_2, y_3)$ satisfies with equation 7.

The surface

$$t = \tau(y_1, y_2, y_3) \tag{14}$$

represents the projection of the cross section of characteristic $y_4 = 0$ by the plane $t = $ constant on the space (y_1, y_2, y_3). This plane separates the undisturbed part of space from the disturbed part and is called a wavefront.

This surface moves in the space (y_1, y_2, y_3) in the course of time. To find the velocity of this motion, denote Δs the displacement of the wavefront in the direction of vector $\nabla \tau$ for time interval Δt. Then the velocity of displacement can be defined as

$$v = \lim_{\Delta t \to 0} \frac{\Delta s}{\Delta t} = \lim_{\Delta t \to 0} \left(\frac{\Delta t}{\Delta s} \right)^{-1}.$$

Replace Δt with $\Delta \tau$ in accordance with equation 14, taking into consideration the relationship following from equation 7,

$$|\nabla \tau| = d\tau / ds = 1/c, \tag{15}$$

where the derivative is taken in the direction of vector $\nabla \tau$. Then

$$v = \lim_{\Delta t \to 0} \left(\frac{\Delta \tau}{\Delta s} \right)^{-1} = \left(\frac{d\tau}{ds} \right)^{-1} = |\nabla \tau|^{-1} = c.$$

Thus, the velocity of wavefront propagation is c.

Let the direction of a tangent to some space curve coincide at each point of this curve with the direction of vector $\nabla\tau$. Such curve is called a ray. Relationship 15, in which the derivation is carried out along the ray, holds true at each point of the ray. Integrating equation 15 along the ray gives

$$\tau = \int_{M_0}^{M} \frac{ds}{c}, \tag{16}$$

where M_0 and $M(y_1, y_2, y_3)$ are fixed and arbitrary points of the ray, respectively. One can see that function τ is the time of propagation of the wavefront from M_0 to M. Function 16 is called the eikonal.

Expression 16 is connected closely with Fermat's principle. In accordance with this principle, the eikonal corresponds to the minimal (or more precisely, extremal) value of the functional 16 along all possible trajectories connecting M_0 and M. A trajectory, satisfying this principle and called the extremal of the functional 16, satisfies Euler's equation

$$\frac{d}{ds}\left(\frac{\mathbf{s}}{c}\right) = \nabla\left(\frac{1}{c}\right), \tag{17}$$

where the derivative is taken in the direction of the unit vector $\mathbf{s}$, tangent to the extremal. Geometric optics shows (e.g., Born and Wolf, 2000; Goldin, 2005) that the extremal coincides with the ray, i.e., the unit vector of the tangent to the ray $\mathbf{s} = \nabla\tau/|\nabla\tau|$ satisfies equation 17.

Propagation of a discontinuity

Now we find the value of the second derivative's discontinuity (equation 11) at the wavefront. Because the nonstationary wave satisfies equation 8, quantity 11 has to satisfy some equation also. To find this equation, differentiate equation 8 with respect to variable y_4, i.e.,

$$\Delta\left(\frac{\partial f^*}{\partial y_4}\right) - 2\nabla\tau \cdot \nabla\left(\frac{\partial^2 f^*}{\partial y_4^2}\right) - \Delta\tau\frac{\partial^2 f^*}{\partial y_4^2} = \frac{\partial F^*}{\partial y_4}, \tag{18}$$

and take to the limit $y_4 \to 0$ in the equation obtained. It follows from expressions 11 and 12 that

$$\lim_{y_4 \to 0} \Delta\left(\frac{\partial f^*}{\partial y_4}\right) = 0, \quad \lim_{y_4 \to 0} \frac{\partial^2 f^*}{\partial y_4^2} = \varphi, \quad \lim_{y_4 \to 0} \frac{\partial F^*}{\partial y_4} = 0.$$

Taking to the limit $y_4 \rightarrow 0$ in equation 18 and accounting for these expressions, obtain the differential equation for the value of discontinuity 11,

$$2\nabla\tau \cdot \nabla\varphi + \varphi\Delta\tau = 0, \tag{19}$$

known as the transport equation.

This equation has the following solution:

$$\varphi = C\exp\left(-\frac{1}{2}\int_0^\tau c^2\Delta\tau\,d\tau\right), \tag{20}$$

where differentiation is carried out along the ray and the arbitrary coefficient C does not depend on τ.

Stationary waves

The mathematical theory of propagation and diffraction of waves usually deals with the case of harmonic oscillations. Nonstationary processes can be reduced to this case by means of the time-frequency Fourier transform.

Writing the right side of wave equation 1 as a Fourier integral,

$$F^*(t) = \frac{1}{2\pi}\int_{-\infty}^{\infty} F(\omega)\exp(-i\omega t)\,d\omega, \quad F(\omega) = \int_{-\infty}^{\infty} F^*(t)\exp(i\omega t)\,dt, \tag{21}$$

look for solution in the form of a Fourier integral

$$f^*(t) = \frac{1}{2\pi}\int_{-\infty}^{\infty} f(\omega)\exp(-i\omega t)\,d\omega. \tag{22}$$

Substituting these expressions into wave equation 1 reduces it to the form

$$\Delta f + k^2 f = F, \quad k = \omega/c. \tag{23}$$

The solution of this equation can be expressed through the solution of the initial nonstationary equation 1 by means of the inverse Fourier transform

$$f(\omega) = \int_{-\infty}^{\infty} f^*(t)\exp(i\omega t)dt. \tag{24}$$

Now write the function $f^*(t)$ in the form 13 so that

$$f(\omega) = \int_{\tau}^{\infty} f^*(y_1,\, y_2,\, y_3,\, t - \tau)\exp(i\omega t)dt. \tag{25}$$

Introduce the new variable of integration $y_4 = t - \tau$. Then

$$f(y_1, y_2, y_3, \omega) = \Phi\exp(i\omega\tau) \tag{26}$$

and

$$\Phi = \int_{0}^{\infty} f^*(y_1, y_2, y_3, y_4)\exp(i\omega y_4)dy_4. \tag{27}$$

Quantities 26 and 27 are a stationary wave and its amplitude, respectively. Substituting equation 26 into 22, we obtain the following representation of the nonstationary wave:

$$f^*(y_1, y_2, y_3, t - \tau) = \frac{1}{2\pi}\int_{-\infty}^{\infty} \Phi\exp(-i\omega y_4)d\omega. \tag{28}$$

The near-front (or high-frequency) approximation

Take the second derivative of integral 28

$$\frac{\partial^2 f^*}{\partial y_4^2} = \frac{1}{2\pi}\int_{-\infty}^{\infty} \Phi''\exp(-i\omega y_4)d\omega, \tag{29}$$

where

$$\Phi'' = (i\omega)^2\Phi. \tag{30}$$

Apply the inverse Fourier transform to the relationship 29 to obtain

$$\Phi'' = \int\limits_0^\infty \frac{\partial^2 f^*}{\partial y_4^2} \exp(i\omega y_4) dy_4. \tag{31}$$

Consider the integral 31 as $\omega \to \infty$. Using the well-known asymptotic formula for Fourier-type integrals (Felsen and Marcuvitz, 1973; Bleistein and Handelsman, 1975),

$$\int\limits_0^\infty \Psi(x) \exp(i\omega x) dx \sim -\Psi(0)/i\omega + O(\omega^{-2}) \quad \text{as} \quad \omega \to \infty, \tag{32}$$

we obtain the asymptotic value of integral 31 in the form

$$\Phi'' = -\frac{1}{i\omega} \left(\frac{\partial^2 f^*}{\partial y_4^2} \right)_{y_4=0} \quad \text{as} \quad \omega \to \infty. \tag{33}$$

Taking into account equation 11, rewrite this expression as

$$\Phi'' = -\varphi/i\omega. \tag{34}$$

Substituting this expression into relationship 30, we obtain

$$\Phi = -\varphi/(i\omega)^3. \tag{35}$$

Substituting equation 20 into 35 and taking into account the fact that the arbitrary coefficient in equation 20 can represent any arbitrary function of frequency $C = C(\omega)$, we obtain the high-frequency approximation of the stationary wave amplitude

$$\Phi = \chi(\omega)G, \quad \chi(\omega) = -C(\omega)/(i\omega)^3 \quad \text{for} \quad \omega \to \infty, \tag{36}$$

$$G = \exp\left(-\frac{1}{2} \int\limits_0^\tau c^2 \Delta\tau \, d\tau \right), \tag{37}$$

where the integration is carried out along the ray and the arbitrary coefficient χ does not depend on τ.

Comparing formulas 37 and 20, it is evident that function 37 satisfies the transport equation

$$2\nabla\tau \cdot \nabla G + G\Delta\tau = 0. \tag{38}$$

Substituting equation 36 into 28, we obtain the near-front approximation of the nonstationary wave

$$f^* = |G| U(t - \tau) \quad \text{as} \quad \omega \to \infty, \tag{39}$$

$$U(t) = \frac{1}{2\pi} \int_{-\infty}^{\infty} \chi(\omega) \exp\left[i(\arg G - \omega t)\right] d\omega. \tag{40}$$

Geometric spreading

This approximate solution to the wave equation can be represented in
geometric terms by using the concept of congruence: "A system of curves
which fills a portion of space in such a way that in general a single curve
passes through each point of the region is called a congruence" (Born and
Wolf, 2000, p. 134).

A two-parameter set of rays can form a congruence. Let α and β be a pair
of parameters specifying an individual ray in a given congruence. Then the
position of any point of this ray can be determined in 3D space by three
parameters (τ, α, β), where τ is the eikonal. The three parameters (τ, α, β)
are called ray coordinates.

A set of rays, described by all possible values of parameters α and β
varying in infinitesimal intervals,

$$\alpha_0 < \alpha < \alpha_0 + d\alpha, \quad \beta_0 < \beta < \beta_0 + d\beta, \tag{41}$$

is called a ray tube.

Take two nearby normal cross sections of the ray tube defined by $\tau =$
constant and $\tau + d\tau =$ constant (Figure 1). Let point $M(\tau, \alpha, \beta)$ of some
ray be determined by Cartesian coordinates (x, y, z) and the other point
$(\tau + d\tau, \alpha + d\alpha, \beta + d\beta)$ by coordinates $x + dx, y + dy, z + dz$. Then
the infinitesimal volume of the ray tube between cross sections is

$$\Delta V = dx\,dy\,dz = D\,d\tau\,d\alpha\,d\beta, \tag{42}$$

where D is the Jacobian of the transition from Cartesian to ray coordinates

$$D = D(x, y, z)/D(\tau, \alpha, \beta) = \begin{vmatrix} x_\tau(\tau, \alpha, \beta) & y_\tau(\tau, \alpha, \beta) & z_\tau(\tau, \alpha, \beta) \\ x_\alpha(\tau, \alpha, \beta) & y_\alpha(\tau, \alpha, \beta) & z_\alpha(\tau, \alpha, \beta) \\ x_\beta(\tau, \alpha, \beta) & y_\beta(\tau, \alpha, \beta) & z_\beta(\tau, \alpha, \beta) \end{vmatrix}. \tag{43}$$

Here the subscripts denote the derivatives with respect to corresponding
variables.

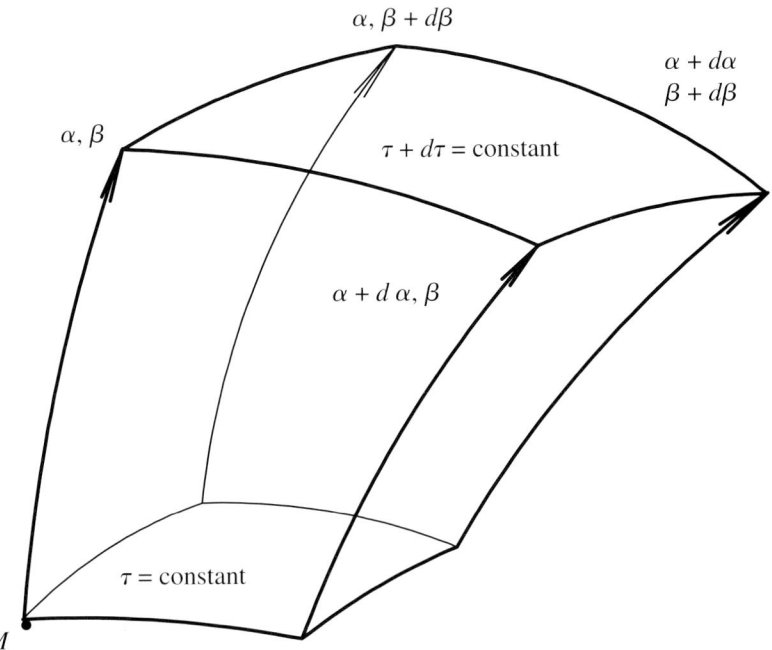

Figure 1. Element of the ray tube between two nearby normal cross sections where M is a current point of the ray.

Now express quantity 43 in terms of the cross section of the ray tube by writing quantity 42 as

$$\Delta V = \Delta \sigma \, ds, \tag{44}$$

where $\Delta \sigma$ is the ray-tube cross-section area at point M and ds is the ray arc length between cross sections at τ and $\tau + d\tau$. By equating expressions 42 and 44, we obtain

$$D = \frac{\Delta \sigma (ds/d\tau)}{d\alpha \, d\beta} = cJ, \quad J = \frac{\Delta \sigma}{d\alpha \, d\beta}. \tag{45}$$

By repeating the above process, it is easy to show that the quantity J corresponds to the Jacobian of the transition from Cartesian coordinates to curvilinear coordinates

$$J = D(x, y, z)/D(s, \alpha, \beta), \tag{46}$$

where s is the arc length along the ray $\alpha = $ constant, $\beta = $ constant.

Quantity J, proportional to the ray-tube cross-section area $\Delta\sigma$ with $d\alpha\,d\beta$ = constant, is the divergence of the ray field or geometric spreading. It characterizes expansion of the ray tube in the wave-propagation process.

The solutions of the wave equation found earlier can be expressed now in terms of these geometric concepts. Because the ray-coordinate system is orthogonal, the differential operations ΔF and ∇F can be written as

$$\Delta F = \sum_{\eta=\tau,\alpha,\beta} \frac{1}{D} \frac{\partial}{\partial\eta} \left(\frac{D}{g_{\eta\eta}} \frac{\partial F}{\partial\eta} \right), \tag{47}$$

$$\nabla F = \sum_{\eta=\tau,\alpha,\beta} \frac{\mathbf{i}_\eta}{\sqrt{g_{\eta\eta}}} \frac{\partial F}{\partial\eta}, \tag{48}$$

$$g_{\eta\eta} = (\partial x/\partial\eta)^2 + (\partial y/\partial\eta)^2 + (\partial z/\partial\eta)^2, \tag{49}$$

where $\mathbf{i}_\eta$ is a unit vector of the corresponding coordinate axis.

Take $F(\tau,\alpha,\beta) = \tau$. Then

$$\Delta\tau = \frac{1}{D} \frac{\partial}{\partial\tau} \left(\frac{D}{g_{\tau\tau}} \right), \tag{50}$$

$$\nabla\tau = \frac{\mathbf{i}_\tau}{\sqrt{g_{\tau\tau}}}. \tag{51}$$

From equation 51, we obtain the scalar product

$$\nabla\tau \cdot \nabla\tau = (\nabla\tau)^2 = 1/g_{\tau\tau}. \tag{52}$$

By equating expressions 7 and 52, we obtain

$$g_{\tau\tau} = c^2. \tag{53}$$

Substituting this expression into equation 50 results in

$$\Delta\tau = \frac{1}{D} \frac{\partial}{\partial\tau} \left(\frac{D}{c^2} \right). \tag{54}$$

Introducing the quantity

$$G = \frac{c}{\sqrt{D}} = \sqrt{\frac{c}{J}}, \tag{55}$$

we rewrite equation 54 as

$$2\partial G/\partial \tau + c^2 G \Delta \tau = 0. \tag{56}$$

It is easy to see that this expression coincides with the transport equation 38. Indeed, from equations 48, 51, and 53, it follows that

$$\nabla_\tau \cdot \nabla G = \left(\frac{\mathbf{i}_\tau}{\sqrt{g_{\tau\tau}}}\right) \cdot \left(\sum_{\eta=\tau,\alpha,\beta} \frac{\mathbf{i}_\eta}{\sqrt{g_{\eta\eta}}} \frac{\partial G}{\partial \eta}\right) = \frac{1}{g_{\tau\tau}} \frac{\partial G}{\partial \tau} (\mathbf{i}_\tau \cdot \mathbf{i}_\tau) = \frac{1}{c^2} \frac{\partial G}{\partial \tau}. \tag{57}$$

Substituting this expression into the transport equation 38, we get equation 56. Thus, the stationary wave amplitude 36 can be expressed through geometric spreading by means of relationship 55.

Substituting equation 55 into 26 and 36, we rewrite the high-frequency approximation of the stationary wave as

$$f = \Phi \exp(i\omega\tau), \quad \Phi = \chi \sqrt{\frac{c}{J}} \quad \text{as} \quad \omega \to \infty. \tag{58}$$

Substituting equation 55 into 39, we rewrite the near-front approximation of the nonstationary wave as

$$f^* = f^*(t - \tau), \quad f^*(t) = \sqrt{\left|\frac{c}{J}\right|}\, U(t) \quad \text{for} \quad t \to \tau, \tag{59}$$

$$U(t) = \frac{1}{2\pi} \int_{-\infty}^{\infty} \chi(\omega)\exp\{-i[(\arg J)/2 + \omega t]\}\, d\omega. \tag{60}$$

The coefficient χ in equation 58 is arbitrary. As a constant of integration of the transport equation 19 or 38, this factor does not depend on τ, although it can depend on the choice of the ray ($\alpha = $ constant, $\beta = $ constant) along which the integration in equation 37 is carried out, i.e.,

$$\chi(\omega) = \chi(\omega, \alpha, \beta). \tag{61}$$

Because this quantity depends on the directions of rays, it has the sense of a directivity pattern of the propagating wave. Because the value of this function is constant for an individual ray, the stationary wave amplitude 36 at an arbitrary point M of the ray can be expressed through its value at any fixed point M_0 of the same ray. Write the amplitude of the wave 58 at the arbitrary points M and M_0 of the same ray,

$$\Phi(M) = \chi(\omega, \alpha, \beta)\sqrt{\frac{c(M)}{J(M)}}, \quad \Phi(M_0) = \chi(\omega, \alpha, \beta)\sqrt{\frac{c(M_0)}{J(M_0)}}, \quad (62)$$

for $\alpha = $ constant, $\beta = $ constant, and eliminate the common factor $\chi(\omega, \alpha, \beta)$. Then

$$\Phi(M) = \Phi(M_0)/L \text{ for } \alpha = \text{constant}, \beta = \text{constant, so that} \quad (63)$$

$$L = \sqrt{\frac{c(M_0)J(M)}{c(M)J(M_0)}}. \quad (64)$$

Thus, the use of equation 58 presupposes that the amplitude value is known at some surface

$$S(M_0) = \text{constant} \quad (65)$$

containing the point M_0. For example, this surface can coincide with the wavefront at some given moment.

The field of a concentrated source

Consider, as an example, a wavefield f^* caused by a nonstationary source in unbounded space. Such a field satisfies wave equation 1

$$\sum_{i=1}^{3} \frac{\partial^2 f^*}{\partial x_i^2} - \frac{1}{c^2}\frac{\partial^2 f^*}{\partial t^2} = F^*, \quad (66)$$

$$F^* = \hat{F}(t)\delta(x - x_0)\delta(y - y_0)\delta(z - z_0), \quad (67)$$

and

$$\hat{F}(t) = 0 \quad \text{for} \quad t < 0, \quad (68)$$

where x_0, y_0, z_0 are the source coordinates, and δ is a delta function determined by the relationships

$$\delta(\zeta) = 0 \quad \text{for} \quad \zeta \neq 0, \quad \int_{-\infty}^{\infty} \delta(\zeta)d\zeta = 1. \tag{69}$$

According to equation 67, the source begins to radiate at the moment $t = 0$. Therefore,

$$f^*(t) = 0 \quad \text{for} \quad t < 0. \tag{70}$$

We consider the retarded solution to wave equation 66 tending to zero at infinity. By representing the time functions f^* and F^* with the Fourier integrals 21 and 22, we can reduce equation 66 to the form 23:

$$\Delta f + k^2 f = F, \quad k = \omega/c, \tag{71}$$

$$F = C(\omega)\delta(x - x_0)\delta(y - y_0)\delta(z - z_0), \tag{72}$$

$$C(\omega) = \int_0^{\infty} \hat{F}(t)\exp(i\omega t)\,dt. \tag{73}$$

Notice that according to the definition of the delta function 69, it is possible to use the following equations instead of equation 71:

$$\left(\Delta + k^2\right)f = 0 \quad \text{for} \quad x \neq x_0, \quad y \neq y_0, \quad z \neq z_0, \tag{74}$$

and

$$\int_V \left(\Delta + k^2\right)f\,dV = C(\omega), \tag{75}$$

where the integration is carried out over the whole space.

The exact solution of equation 71 in a case of an arbitrary inhomogeneous medium is not known. However, the high-frequency approximation of such a solution can be found by using expressions 58 and 63. To do this, it is sufficient to find quantities $J(M_0)$ and $\Phi(M_0)$ at the surface 65.

Let the surface 65 be a sphere with the center at point O. Let the radius r of this sphere be so small that we can neglect the variation of wave velocity within the sphere

$$c \approx c(O) = c(M_0). \tag{76}$$

Then the eikonal value at the sphere will be determined by equation 16 as follows:

$$\tau(M_0) = \int_O^{M_0} \frac{ds}{c(O)} = \frac{1}{c(O)} \int_O^{M_0} ds = r/c(O) = r/c(M_0). \tag{77}$$

Find the geometric spreading at the sphere $J(M_0)$. Let the coordinate surfaces $\alpha = $ constant and $\beta = $ constant be mutually orthogonal. Then the normal cross section of the ray tube can be written as

$$\Delta\sigma = r_\alpha d\alpha' \cdot r_\beta d\beta', \tag{78}$$

where r_α and r_β are curvature radii of cross sections of the surface $\tau = $ constant by surfaces $\alpha = $ constant and $\beta = $ constant; $d\alpha'$ and $d\beta'$ are the angular sizes of these cross sections, respectively (see Figure 2). Substituting this expression into equation 45, we obtain

$$J = r_\alpha r_\beta \frac{d\alpha'}{d\alpha} \frac{d\beta'}{d\beta}. \tag{79}$$

Because of condition 76, the rays are locally rectilinear, and the following relationships hold true:

$$d\alpha' = d\alpha, \quad d\beta' = d\beta, \quad r_\alpha = r_\beta = r. \tag{80}$$

Substituting these expressions into equation 79, we obtain

$$J(M_0) = r^2. \tag{81}$$

To find the quantity $\Phi(M_0)$, it is necessary to know the solution of equation 71 on sphere 65. It is easy to check that under condition 76, the following function satisfies this equation:

$$f = C' \frac{\exp(ikr)}{r}, \tag{82}$$

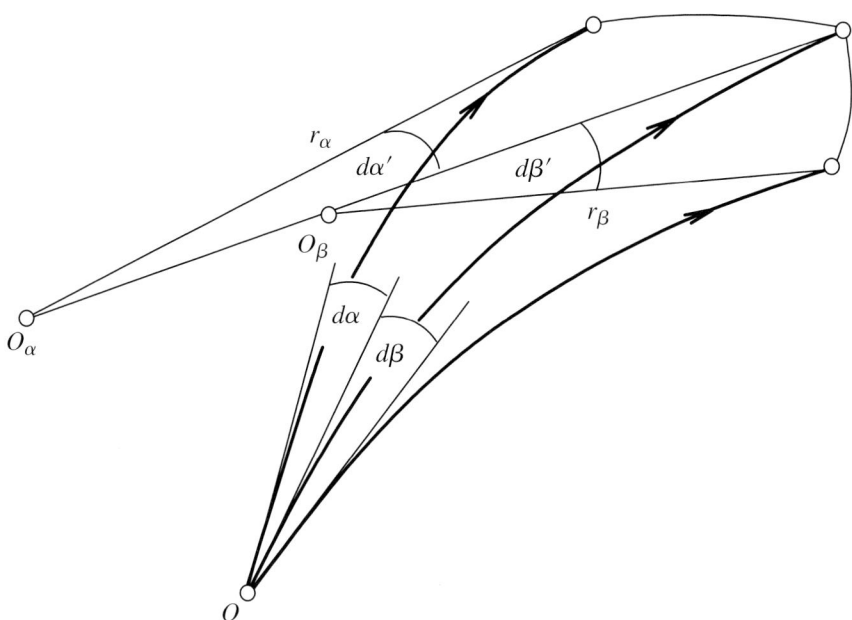

Figure 2. Radii (r_α and r_β) and centers (O_α and O_β) of curvature of wavefront cross sections.

where the unknown constant coefficient C' can be found by means of condition 75. In addition, on the strength of relationship 74, it is sufficient to integrate over the volume bounded by the small sphere as $r \to 0$.

To find the value of the integral, transfer to spherical coordinates

$$x = r\sin\varphi\,\cos\theta, \quad y = r\sin\varphi\,\sin\theta, \quad z = r\cos\varphi,$$

$$\text{for} \quad -\pi \le \theta \le \pi, \quad 0 \le \varphi \le \pi. \tag{83}$$

The elementary volume is then

$$dV = r^2 \sin\varphi\,dr\,d\varphi\,d\theta, \tag{84}$$

and condition 75 can be written as

$$I(f) = C(\omega), \tag{85}$$

where

$$I(f) = \int_{-\pi}^{\pi}\int_{0}^{\pi}\int_{0}^{r}\left[\left(\Delta + k^2\right)f\right]r^2 \sin\varphi\,dr\,d\varphi\,d\theta. \tag{86}$$

The direct substitution of equation 82 into equation 86 makes no sense because the integrand equals zero with $r \neq 0$. Consider expressions 82 and 85 in the following limits:

$$f = \lim_{\varepsilon \to 0} f_\varepsilon, \quad f_\varepsilon = C' \frac{\exp[ik(r + \varepsilon)]}{r + \varepsilon}, \tag{87}$$

$$\lim_{\varepsilon \to 0} I(f_\varepsilon) = C(\omega). \tag{88}$$

Substituting f_ε into the differential operator in equation 74, we obtain

$$(\Delta + k^2) f_\varepsilon = -\frac{2\varepsilon[1 + ik(r + \varepsilon)]}{r(r + \varepsilon)^2} f_\varepsilon. \tag{89}$$

Because $r \to 0$ and $\varepsilon \to 0$, it is possible to neglect small quantities that do not influence the singularity of the integrand. Then

$$(\Delta + k^2) f_\varepsilon \approx -\frac{2\varepsilon C'}{r(r + \varepsilon)^3}. \tag{90}$$

Introducing this expression into integral 86, we obtain

$$I(f_\varepsilon) = -2\varepsilon C' I_r I_\varphi I_\theta,$$

$$I_r = \int_0^r \frac{r \, dr}{(r + \varepsilon)^3} = \left[-\frac{1}{r + \varepsilon} + \frac{\varepsilon}{2(r + \varepsilon)^2} \right]\Bigg|_0^r$$

$$= -\frac{1}{r + \varepsilon} + \frac{\varepsilon}{2(r + \varepsilon)^2} + \frac{1}{2\varepsilon},$$

$$I_\varphi = \int_0^\pi \sin\varphi \, d\varphi = (-\cos\varphi)\big|_0^\pi = 2, \quad I_\theta = \int_{-\pi}^\pi d\theta = 2\pi \tag{91}$$

or

$$I(f_\varepsilon) = -4\pi C' \left[1 + \frac{2\varepsilon}{r + \varepsilon} - \frac{\varepsilon^2}{2(r + \varepsilon)^2} \right]. \tag{92}$$

Substituting this expression into equation 88, we obtain the unknown coefficient

$$C' = -C(\omega)/4\pi. \tag{93}$$

Substituting this expression into equation 82, we obtain

$$f(M_0) = -\frac{C(\omega)\exp(ikr)}{4\pi r}. \tag{94}$$

Taking into consideration relationship 77, we rewrite this expression in the form of a stationary wave

$$f(M_0) = \Phi(M_0)\exp\left[i\omega\tau(M_0)\right], \tag{95}$$

$$\Phi(M_0) = -\frac{C(\omega)}{4\pi r}. \tag{96}$$

This expression gives the value of the amplitude at the surface 65. Substituting equations 96 and 81 into 63, we obtain the stationary wave amplitude outside this surface as

$$\Phi(M) = -\frac{C(\omega)}{4\pi}\sqrt{\frac{c(M)}{c(O)J(M)}}. \tag{97}$$

It follows from equations 58 and 97 that

$$\chi = -\frac{C(\omega)}{4\pi\sqrt{c(O)}}. \tag{98}$$

Introducing this expression into equation 60, the pulse form of the non-stationary wave is given as

$$U(t) = -\frac{1}{8\pi^2\,c(O)}\int_{-\infty}^{\infty} C(\omega)\exp\left\{-i\left[(\arg J)/2 + \omega t\right]\right\}d\omega. \tag{99}$$

Limitations on the elementary theory

Expression 63 is derived under the assumption that the directivity pattern 61 exists. Regions of space where the directivity pattern does not exist (and therefore expression 58 fails) are called diffraction boundary layers.

Another limitation on the use of expression 58 is connected with the existence of ray-congruence envelopes called caustics, where the geometric spreading J becomes zero. The transport equation 38 fails when nearing caustics because its solution tends to infinity as $J \to 0$. The cause of this can be traced to the process of deriving transport equation 19. Indeed, we obtained equation 19 by deriving wave equation 8 with respect to y_4 and then assuming that limit 11 exists. However, equation 55 shows that the latter assumption fails at the surface $J = 0$. Therefore, equation 58 fails in the neighborhood of caustics. However, the expression holds true outside such neighborhoods for a wave crossing any number of caustics in the process of its propagation.

The above considerations can be repeated for the case of nonstationary wave 59. In this case, quantity 61 must be given as a function of frequency and two ray coordinates. This is possible only outside diffraction boundary layers. Expression 59 holds true only outside the neighborhoods of caustics, which are taken into consideration in equation 60 by means of the quantity arg J. Function 60 does not change its character in wave propagation at the parts of the ray where condition arg J = constant holds true. When crossing over a caustic, quantity J changes its sign, and therefore, quantity arg J in integral 60 changes intermittently by π. The quantity under the exponent in equation 60 being stored in the propagation process resulting from changes of $(\arg J)/2$ is called the caustic index. In the neighborhood of a concentrated source of oscillations, where $J > 0$, i.e., arg $J = 0$, the pulse form is determined completely by its frequency spectrum 61. In wave propagation, the pulse form (equation 99) can vary only because of variation in the caustic index.

Elastic waves in isotropic media

Longitudinal and shear waves

The above elementary theory can be applied to the elastodynamic equations of motion in isotropic media

$$(\lambda + \mu)\nabla \operatorname{div} \mathbf{u}^* + \mu\Delta\mathbf{u}^* + \nabla\lambda \operatorname{div} \mathbf{u}^*$$

$$+ \nabla\mu \times \operatorname{rot} \mathbf{u}^* + 2(\nabla\mu \cdot \nabla)\mathbf{u}^* = \rho\frac{\partial^2\mathbf{u}^*}{\partial t^2}, \tag{100}$$

where $\mathbf{u}^*$ is the particle displacement vector; λ, μ, and ρ are, respectively, Lamé's parameters and density, which are functions of the space coordinates

(x_1, x_2, x_3). The following notations of differential operations are used here:

$$\Delta \mathbf{u}^* = \nabla \operatorname{div} \mathbf{u}^* - \operatorname{rotrot} \mathbf{u}^*, \quad (\nabla \mu \cdot \nabla) = \sum_{i=1}^{3} \mathbf{i}_i \frac{\partial \mu}{\partial x_i} \frac{\partial}{\partial x_i}, \qquad (101)$$

where $\mathbf{i}_i$ is the unit vector of the ith coordinate axis.

Look for a solution in the form of a Fourier integral

$$\mathbf{u}^*(t) = \frac{1}{2\pi} \int_{-\infty}^{\infty} \mathbf{u}(\omega) \exp(-i\omega t) d\omega, \qquad (102)$$

with inverse

$$\mathbf{u}(\omega) = \int_{-\infty}^{\infty} \mathbf{u}^*(t) \exp(i\omega t) dt, \qquad (103)$$

where (for brevity) dependence of the displacement vector on space coordinates is not shown.

When $\omega \to \infty$, the solution of equation 100 admits the representation (e.g., Červený, 2001)

$$\mathbf{u} = (\mathbf{u}_p + \mathbf{u}_s)\left[1 + O(\omega^{-\varepsilon})\right], \quad \varepsilon > 0, \quad \text{as} \quad \omega \to \infty, \qquad (104)$$

$$\mathbf{u}_p = \nabla\left(\frac{\phi_p}{\sqrt{\rho}}\right), \quad \mathbf{u}_s = \operatorname{rot}\left(\frac{\phi_s \mathbf{\Psi}}{\sqrt{\rho}}\right), \qquad (105)$$

where scalar functions ϕ_p and ϕ_s satisfy the stationary wave equations

$$\Delta \phi_n + k_n^2 \phi_n = 0, \quad k_n = \omega/c_n \quad \text{for} \quad n = p, s \qquad (106)$$

and

$$c_p = \sqrt{(\lambda + 2\mu)/\rho}, \quad c_s = \sqrt{\mu/\rho}, \qquad (107)$$

and the vector $\mathbf{\Psi}$ does not depend on ω (the explicit expression of this vector will be given below).

Because we have considered solutions of such equations with $\omega \to \infty$, we immediately can write the high-frequency solutions of equations 106, using expression 58, as

$$\phi_n = \chi'_n \sqrt{\frac{c_n}{J_n}} \exp(i\omega\tau_n) \quad \text{for} \quad \omega \to \infty; \quad n = p, s, \qquad (108)$$

where τ_n and J_n are the eikonal and geometric spreading of the stationary wave, determined at the ray congruence (α_n, β_n) and propagating with velocity c_n. The arbitrary function of frequency χ'_n is constant at the ray $\alpha_n = $ constant, $\beta_n = $ constant. By using this expression, we can simplify formula 105.

Substituting equation 108 into expressions 105 and using notations

$$\varphi_n = \Phi'_n \exp(i\omega\tau_n), \quad \Phi'_n = \chi'_n \sqrt{\frac{c_n}{\rho J_n}} \quad \text{for} \quad n = p, s, \quad (109)$$

$$\mathbf{s}_n = \nabla\tau_n / |\nabla\tau_n| \quad \text{for} \quad n = p, s, \qquad (110)$$

we obtain

$$\mathbf{u}_p = \nabla\varphi_p = \left(\nabla\Phi'_p\right)\exp\left(i\omega\tau_p\right) + \Phi'_p\nabla\exp\left(i\omega\tau_p\right)$$

$$= \left(\nabla\Phi'_p + i\omega\Phi'_p\,\nabla\tau_p\right)\exp\left(i\omega\tau_p\right) \approx ik_p\Phi'_p\,\mathbf{s}_p\exp\left(i\omega\tau_p\right) \quad \text{as} \quad \omega \to \infty$$

$$(111)$$

and

$$\mathbf{u}_s = \text{rot}(\varphi_s\mathbf{\Psi}) = \varphi_s\,\text{rot}\,\mathbf{\Psi} + \nabla\varphi_s \times \mathbf{\Psi}$$

$$= \left[\Phi'_s\,\text{rot}\,\mathbf{\Psi} + \left(\nabla\Phi'_s + ik_s\Phi'_s\mathbf{s}_s\right) \times \mathbf{\Psi}\right]\exp\left(i\omega\tau_s\right)$$

$$\approx ik_s\Phi'_s\,\mathbf{e}_s\exp\left(i\omega\tau_s\right) \quad \text{as} \quad \omega \to \infty, \qquad (112)$$

where

$$\mathbf{e}_s = \mathbf{s}_s \times \mathbf{\Psi}. \qquad (113)$$

The vector $\mathbf{\Psi}$ has the following representation (Červený, 2001):

$$\mathbf{\Psi} = C\mathbf{s}_s - \mathbf{n}\sin\theta - \mathbf{b}\cos\theta, \qquad (114)$$

where C is an arbitrary constant $\mathbf{s}_s$, $\mathbf{n}$, $\mathbf{b}$ are the unit vectors of the tangent, principal normal, and binormal, respectively, to the ray, along which the stationary wave 108 propagates with velocity c_s. Quantity θ can be expressed in terms of the torsion T of this ray as follows:

$$\theta(\tau_s) = \int_0^{\tau_s} c_s T d\tau_s + \text{constant}, \tag{115}$$

where the integration is carried out along the ray.

Substituting equation 114 into 113 and using the usual rule for the vector product in the coordinate basis $(\mathbf{s}_s, \mathbf{n}, \mathbf{b})$, we obtain

$$\mathbf{e}_s = \mathbf{s}_s(1, 0, 0) \times \mathbf{\Psi}(C, -\sin\theta, -\cos\theta) = \mathbf{n}\cos\theta - \mathbf{b}\sin\theta. \tag{116}$$

Now write equation 105 as

$$\mathbf{u}_n = u_n\mathbf{e}_n, \quad u_n = \Phi_n\exp(i\omega\tau_n), \quad \Phi_n = \frac{\chi_n}{\sqrt{\rho c_n J_n}},$$

$$\text{as} \quad \omega \to \infty, \quad n = p, s, \tag{117}$$

$$\mathbf{e}_p = \mathbf{s}_p, \quad \mathbf{e}_s = \mathbf{n}\cos\theta - \mathbf{b}\sin\theta, \tag{118}$$

where $\chi_n = i\omega\chi'_n$.

The section above titled "The near-front (or high-frequency) approximation" shows that a solution of equations of type 106 as $\omega \to \infty$ corresponds to a solution of the nonstationary wave equation in the neighborhood of the wavefront $t \to \tau_n$. Therefore, substituting equation 117 into integral 102, we obtain solutions of the nonstationary equation 100 in the neighborhoods of the wavefronts

$$\mathbf{u}_n^* = u_n\mathbf{e}_n, \quad u_n = \frac{U_n(t - \tau_n)}{\sqrt{|\rho c_n J_n|}} \quad \text{as} \quad t \to \tau_n, \quad n = p, s \tag{119}$$

and

$$U_n(t) = \frac{1}{2\pi} \int_{-\infty}^{\infty} \chi_n\exp\left\{-i\left[(\arg J_n)/2 + \omega t\right]\right\}d\omega. \tag{120}$$

Thus, equation 100 has two types of solutions in the form of nonstationary waves called longitudinal ($n = p$) and shear ($n = s$) waves. The displacement vector of longitudinal wave 118 is directed along the tangent to the ray. For the shear wave, it is directed perpendicular to ray 116.

Law of conservation of the energy flux

Consider a relationship making clear the physical sense of the high-frequency (or near-front) approximation of elastic waves (see, e.g., Achenbach, 1973). To begin with, remember the concept of a flux. The flux of a nonstationary scalar field $F(x, y, z, t)$ through the small oriented area element $\Delta\sigma$ with the unit normal vector $\mathbf{s}$ can be determined as

$$\boldsymbol{\varepsilon} = \mathbf{v}\Delta\sigma\frac{\partial F}{\partial s}, \quad \mathbf{v} = \frac{ds}{dt}\,\mathbf{s}, \tag{121}$$

in which ds (or ∂s) is the differential of the arc length of a curve tangential to the vector $\mathbf{s}$. Quantities $\partial F/\partial s$ represent field density in the direction of the vector $\mathbf{s}$, and $\mathbf{v}$ is the vector of field transport velocity in the same direction.

Use the above-mentioned definitions to introduce the energy flux of the nonstationary wave 119 in the direction of the ray. In this case, it is necessary to take the energy as field F and the energy density as $\partial F/\partial s$. The vector of transport velocity has to be taken as $\mathbf{v} = c_n\mathbf{s}_n$, where c_n and $\mathbf{s}_n$ are determined by equations 107 and 110. Then quantity $\Delta\sigma$ corresponds to the area of the wavefront element from equation 45, namely $\Delta\sigma = J_n\,d\alpha_n\,d\beta_n$, where J_n is the geometric spreading and α_n and β_n are ray coordinates. Substituting all of these quantities into equation 121, we obtain the energy flux of the nonstationary wave,

$$\boldsymbol{\varepsilon} = \varepsilon\,\mathbf{s}_n, \quad \varepsilon = c_n J_n E_n\,d\alpha_n\,d\beta_n, \quad n = p,\, s. \tag{122}$$

Here E_n is the energy density that can be written as

$$E_n = \frac{\rho}{2}\left|\frac{\partial u_n^*}{\partial t}\right|^2 + \hat{E}_n, \tag{123}$$

where the first term corresponds to kinetic energy density and the second to potential energy density. It is a well-known fact of elastodynamics that $\hat{E}_n$ is a homogeneous quadratic form of the stress tensor components. However, there is no need for an explicit expression for this function in the following.

Now consider the case of harmonic oscillations:

$$u_n^* = \frac{1}{2\pi}|u_n|\cos(\arg u_n - \omega t). \tag{124}$$

One can see that this expression corresponds to the scalar harmonic of the Fourier integral 102 for the fixed frequency ω. It is also a well-known fact of elastodynamics that in the case of harmonic oscillations, the densities of kinetic and potential energies are equal (Hönl et al., 1961). Therefore, equation 123 can be rewritten as

$$E_n = \rho \left|\frac{\partial u_n^*}{\partial t}\right|^2, \tag{125}$$

where u_n^* is determined by equation 124.

Instead of equation 125, it is more convenient to consider the average energy density over one period of oscillation:

$$\langle E_n \rangle = \frac{\omega}{2\pi} \int_0^{2\pi/\omega} E_n(t)\, dt. \tag{126}$$

Substituting equation 124 and 125 into 126, we obtain

$$\langle E_n \rangle = \rho|u_n|^2 \left(\frac{\omega}{2\pi}\right)^3 I, \quad I = \int_0^{2\pi/\omega} \sin^2(\arg u_n - \omega t)\, dt = \pi/\omega.$$

Substituting u_n from equation 117 into these formulas, we obtain

$$\langle E_n \rangle = \frac{1}{2c_n J_n}\left(\frac{\omega \chi_n}{2\pi}\right)^2. \tag{127}$$

Using this expression instead of the quantity E_n in equation 122, we obtain the average energy flux over one period of oscillation:

$$\langle \varepsilon \rangle = \langle \varepsilon \rangle \mathbf{s}_n, \quad \langle \varepsilon \rangle = \left(\frac{\omega \chi_n}{2\pi}\right)^2 \frac{d\alpha_n d\beta_n}{2}. \tag{128}$$

Equation 128 shows that the energy flux is constant in any normal cross section of the ray tube. Thus the asymptotic formulas of elementary wave-propagation theory describe only those components of wavefields that comply with the energy-flux conservation law.

Reflected/transmitted waves

Consideration at interfaces

Propagation of waves in media formed by combinations of regions and interfaces can be described in the framework of elementary wave theory. Functions describing physical properties within regions are continuous and are changing slowly. A surface formed by points of discontinuity of any of these functions is called an interface. Encircle an arbitrary point on the interface with a sphere of small radius. A ball bounded by such a sphere forms a neighborhood of the interface point. This neighborhood is divided into two semineighborhoods. The interface point is considered to be regular if the interface is continuous here, along with its first and second tangential derivatives, and if the functions describing the media's physical properties are continuous in each semineighborhood. A part of the interface is considered smooth if its points are all regular. Here we will consider the case of smooth interfaces.

A wavefield in each individual region must satisfy the corresponding equation of motion and some additional physical conditions that provide the unique solution. The latter include conditions at the source of oscillations describing the initial wave and interface conditions that usually require a smooth and continuous solution on crossing the interface. If the regions expand to infinity, it is necessary to include conditions at infinity. Because expressions 58 and 59 satisfy the equations of motion, it is sufficient in the elementary theory framework to find ray congruences and coefficients that satisfy the additional conditions only.

Take the case of a single interface. Suppose conditions at the source are given in the form of the directivity pattern 61 at some surface 65 encircling the source. Then we can find the initial wave emitted by the source, constructing the corresponding congruence of rays and using the given function 61 along with equation 63. To satisfy the interface conditions, we must consider the initial wave as the incident wave. We also introduce some additional waves — the reflected wave in the region where the incident wave is given and the transmitted wave on the other side of the interface. We will consider the problem of finding expressions for these waves in the case of the stationary wave equation.

Local plane-interface approximation

Let the expression

$$f_m = \boldsymbol{\Phi}_m \exp(i\omega\tau_m) \tag{129}$$

represent the incident ($m = 0$), reflected ($m = 1$), or transmitted ($m = 2$) wave. In accordance with equation 63, the amplitude of any of these waves at an arbitrary point M of any ray can be expressed through its value at point S at the interface as

$$\Phi_m(M) = \frac{\Phi_m(S)}{L_m}, \quad L_m = \sqrt{\frac{c_m(S)J_m(M)}{c_m(M)J_m(S)}}. \tag{130}$$

Therefore, the problem of finding reflected/transmitted wave amplitudes consists of finding the quantities $\Phi_m(S)$ with $m = 1, 2$. In the vicinity of the point of incidence, it is possible to neglect the inhomogeneity of the media and curvature of the interface and wavefronts. Then the problem can be reduced to considering the reflection/transmission of plane waves at the plane interface separating homogeneous media. We consider this well-known problem, changing its traditional formulation. Orientation of the interface at the point of incidence usually is determined by the position of the normal to the interface. Instead, we will determine the orientation of the interface by the position of its tangent plane.

Determine the tangent plane to the interface by a pair of straight lines — guides L and L', intersecting at the point of incidence S. Let (x_1, x_2, x_3) be Cartesian coordinates with the plane $x_3 = 0$ coinciding with the tangent plane (L, L') and with the origin at the point of incidence.

When $x_1^2 + x_2^2 + x_3^2 \ll 1$, fix the wave amplitude at the point of incidence

$$\Phi_m(x_1, x_2, x_3) = \Phi_m(0, 0, 0) \tag{131}$$

and approximate the eikonal by the linear part of its Taylor expansion

$$\tau_m(x_1, x_2, x_3) = \tau_m(0, 0, 0) + \sum_{j=1}^{3} x_j (\partial \tau_m / \partial x_j)_{x_1=x_2=x_3=0}. \tag{132}$$

Use the formula for a directional derivative

$$\frac{\partial \tau_m}{\partial \lambda} = |\nabla \tau_m| \cos(\nabla \tau_m \cdot \mathbf{i}_\lambda) = \frac{\cos q}{c_m}, \tag{133}$$

in which $\mathbf{i}_\lambda$ is the unit vector along the direction of derivation, and q is the angle between vectors $\nabla \tau_m$ and $\mathbf{i}_\lambda$. Then

$$\left(\frac{\partial \tau_m}{\partial x_j} \right)_{x_1=x_2=x_3=0} = \frac{\cos q_{jm}}{c_m}, \tag{134}$$

in which q_{jm} is the angle between vector $\nabla \tau_m$ and axis x_j.

Substitute equations 131 through 134 into equation 129 and write the local plane-interface approximation of the incident and reflected/transmitted waves in the neighborhood of point S in the form

$$f_m = f_0(S)F_m, \tag{135}$$

$$F_m = K_m \exp\left[i(k_{1m}x_1 + k_{2m}x_2 + k_{3m}x_3)\right], \quad K_0 = 1, \tag{136}$$

$$k_{jm} = k_m \cos q_{jm} \quad \text{for} \quad j = 1, 2, 3, \quad k_m = \omega/c_m, \tag{137}$$

and

$$\cos^2 q_{1m} + \cos^2 q_{2m} + \cos^2 q_{3m} = 1, \tag{138}$$

in which q_{jm} is the angle between the wave vector $\mathbf{k}_m(k_{1m}, k_{2m}, k_{3m})$ and axis x_j. The unknown cofactors K_m are called the reflection ($m = 1$) and transmission ($m = 2$) coefficients.

The reflection/transmission problem

We consider as an example the acoustic problem in which f_m corresponds to a sound pressure wave and quantity $\nabla f_m/i\omega\rho_m$ (ρ_m is the medium density) corresponds to the particle velocity of this wave. In this case, the interface conditions usually express the continuity of sound pressure and continuity of the particle velocity component normal to the interface.

Then the reflection/transmission problem turns to the integration of two Helmholtz equations,

$$\left(\nabla^2 + k_m^2\right)F_m = 0, \quad m = 1, 2, \tag{139}$$

under the following interface conditions:

$$F_0 + F_1 = F_2, \quad \frac{1}{\rho_1}\frac{\partial}{\partial x_3}(F_0 + F_1) = \frac{1}{\rho_2}\frac{\partial F_2}{\partial x_3} \quad \text{with} \quad x_3 = 0. \tag{140}$$

Equations 139 are satisfied because we look for plane-wave solutions as in equation 135.

To find the reflection/transmission coefficients K_m and angles q_{jm}, substitute equations 136 into conditions 140:

$$\left(1 + K_1\right)\exp\left(i\ell_1\right) = K_2\exp\left(i\ell_2\right),$$
$$a_{31}\left(K_1 - 1\right)\exp\left(i\ell_1\right) = -a_{32}K_2\exp\left(i\ell_2\right), \tag{141}$$

where

$$a_{jm} = \frac{\cos q_{jm}}{c_m\rho_m}, \quad \ell_m = \frac{\omega}{c_m}(x_1\cos q_{1m} + x_2\cos q_{2m}). \tag{142}$$

We must formulate conditions that make it possible to solve this problem. To do this, we express the direction of $\mathbf{k}_m$ by its angles with respect to lines L and L'.

Let axis x_1 coincide with L (Figure 3). Then the angle between $\mathbf{k}_m$ and L is q_{1m}. Let q'_{1m} be the angle between $\mathbf{k}_m$ and L'. Then

$$\cos q'_{1m} = \cos q_{1m}\cos \zeta + \cos q_{2m}\sin \zeta, \tag{143}$$

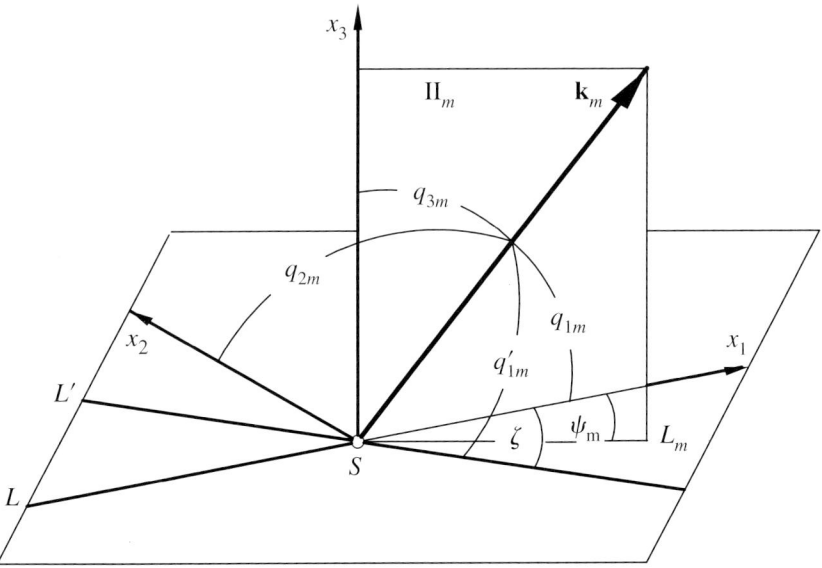

Figure 3. Definition of wave vector $\mathbf{k}_m$ by its angles with respect to lines L and L' on the plane interface (Klem-Musatov, 1994).

in which ζ is the angle between L and L'. From equation 143, it follows that

$$\cos q_{2m} = \left(\cos q'_{1m} - \cos q_{1m} \cos \zeta\right)/\sin \zeta. \tag{144}$$

We always can choose the direction of axis x_1 so that $q_{1m} \leq \pi/2$, $q'_{1m} \leq \pi/2$. From equations 138 and 144, it follows that

$$\cos q_{3m} = \sqrt{\sin^2 q_{1m} - \left(\cos q'_{1m} - \cos q_{1m} \cos \zeta\right)^2/\sin^2 \zeta}. \tag{145}$$

Equations 144 and 145 express the direction of vector $\mathbf{k}_m$ through its angles with lines L and L'.

Reflection/transmission coefficients can be found under conditions

$$\frac{\cos q_{1m}}{c_m} = \frac{\cos q_{10}}{c_0}, \tag{146}$$

$$\frac{\cos q'_{1m}}{c_m} = \frac{\cos q'_{10}}{c_0}. \tag{147}$$

Indeed, by substituting equations 146 and 147 into equation 144 and taking into account the relationship followed from equation 146,

$$\cos q_{20} = \left(\cos q'_{10} - \cos q_{10} \cos \zeta\right)/\sin \zeta,$$

we have

$$\cos q_{2m} = \frac{c_m}{c_0} \cos q_{20}. \tag{148}$$

Substituting equations 146 and 148 into equation 142, we have $\ell_m = \ell_0$ with $m = 1, 2$. Equations 141 become the algebraic equations

$$K_1 - K_2 = -1, \quad a_{31}K_1 + a_{32}K_2 = a_{31}, \tag{149}$$

and we obtain the well-known expressions for reflection/transmission coefficients

$$K_1 = \frac{a_{31} - a_{32}}{a_{31} + a_{32}}, \quad K_2 = \frac{2a_{31}}{a_{31} + a_{32}}. \tag{150}$$

Reflected/transmitted rays

Equations 146 and 147 determine the directions of reflected/transmitted rays. Equation 146 puts a limitation on possible directions of wave vector $\mathbf{k}_m$. All allowable directions form a cone with its vertex at the point of incidence. Its axis is line L. Its apex angle is $2q_{1m}$. Condition 147 produces the second cone of allowable directions with the axis L'. Because conditions 146 and 147 both must be satisfied, vector $\mathbf{k}_m$ must belong to both cones (Figure 4). Then it belongs to the line of intersection of those cones. Thus the incident $(m = 0)$, reflected $(m = 1)$, and transmitted $(m = 2)$ rays are determined by angles q_{1m} and q'_{1m}.

Snell's law

The rule determining the directions of reflected/transmitted rays can be rewritten in a more traditional form. Wave vector $\mathbf{k}_m$ lies in the plane Π_m containing the normal x_3 to the interface. Find the dihedral angle ψ_m between the planes Π_m and $x_2 = 0$ (Figure 3). Let L_m be a line of intersection of planes Π_m and $x_3 = 0$. The cosine of the angle between $\mathbf{k}_m$ and the line of intersection of planes Π_m and $x_3 = 0$ is

$$\cos(\pi/2 - q_{3m}) = \cos q_{1m} \cos \psi_m + \cos q_{2m} \sin \psi_m. \qquad (151)$$

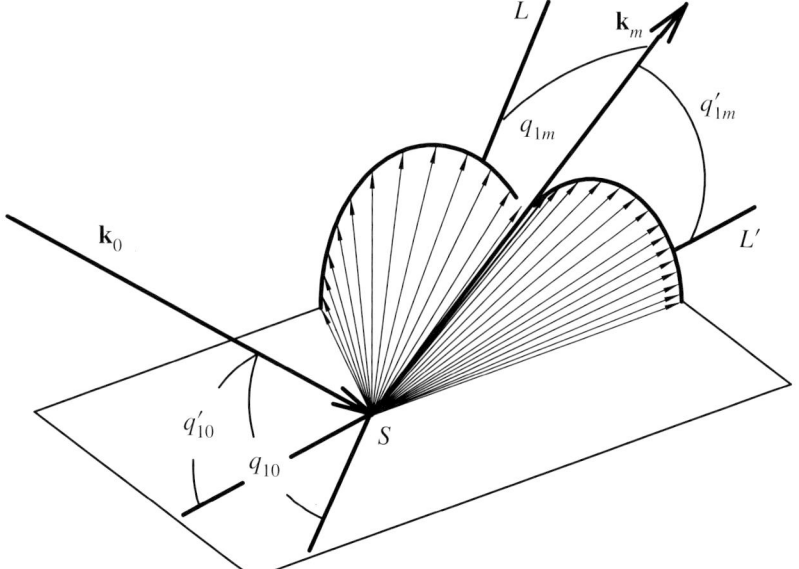

Figure 4. Definition of the reflected/transmitted wave vector $\mathbf{k}_m$ by two cones of permitted directions. The axes of cones coincide with lines L and L'.

From equation 145, we obtain

$$\sin q_{3m} = \sqrt{\cos^2 q_{1m} + (\cos q'_{1m} - \cos q_{1m} \cos \zeta)^2 / \sin^2 \zeta}. \quad (152)$$

By substituting equations 146 and 147 into equation 152 and taking into account that

$$\sin q_{30} = \sqrt{\cos^2 q_{10} - (\cos q'_{10} - \cos q_{10} \cos \zeta)^2 / \sin^2 \zeta},$$

we obtain

$$\sin q_{3m} = \frac{c_m}{c_0} \sin q_{30}. \quad (153)$$

By substituting equations 146, 148, and 153 into equation 151, we obtain

$$\sin q_{30} = \cos q_{10} \cos \psi_m + \cos q_{20} \sin \psi_m, \quad (154)$$

and from equation 154, it follows that the dihedral angles ψ_m satisfy

$$\psi_m = \psi_0 \quad \text{with} \quad m = 1, 2 \quad (155)$$

such that equations 153 and 155 express Snell's law.

Chapter 2

Edge Waves as a Solution
of the Wave Equation

Edge diffraction

The reflection/transmission problem
at an irregular interface

A line formed by points of discontinuity of the interface or its first tangential derivatives is called an edge. A point of the edge is considered to be regular if the corresponding line is continuous along with its first tangential derivative. The edge is considered smooth if its points are all regular.

In the near vicinity of any point of a smooth edge, we can neglect its curvature and approximate the interface by two half planes touching at the edge (Figure 1). Then reflected/transmitted wavefields must satisfy the boundary conditions at both half planes. By using cylindrical coordinates (r, θ, z) where axis z coincides with the edge, we can write the boundary conditions as

$$F_0 + F_1 = F_2, \quad \frac{1}{\rho_1} \frac{\partial}{\partial \theta}(F_0 + F_1) = \frac{1}{\rho_2} \frac{\partial F_2}{\partial \theta}$$

$$\text{for} \quad \theta = \theta_\ell \quad \text{for} \quad \ell = 1, 2, \tag{1}$$

where θ_ℓ is the coordinate of the ℓth half plane. Then the reflection/transmission problem reduces to the integration of equations 139 of Chapter 1 under conditions 1 above.

Also rewrite equation 136 of Chapter 1 in cylindrical coordinates. Let points $x_1 = x_2 = x_3 = 0$ and $r = z = 0$ coincide, axes x_1 and z coincide, and planes $x_2 = 0$ and $\theta = 0$ coincide. Then

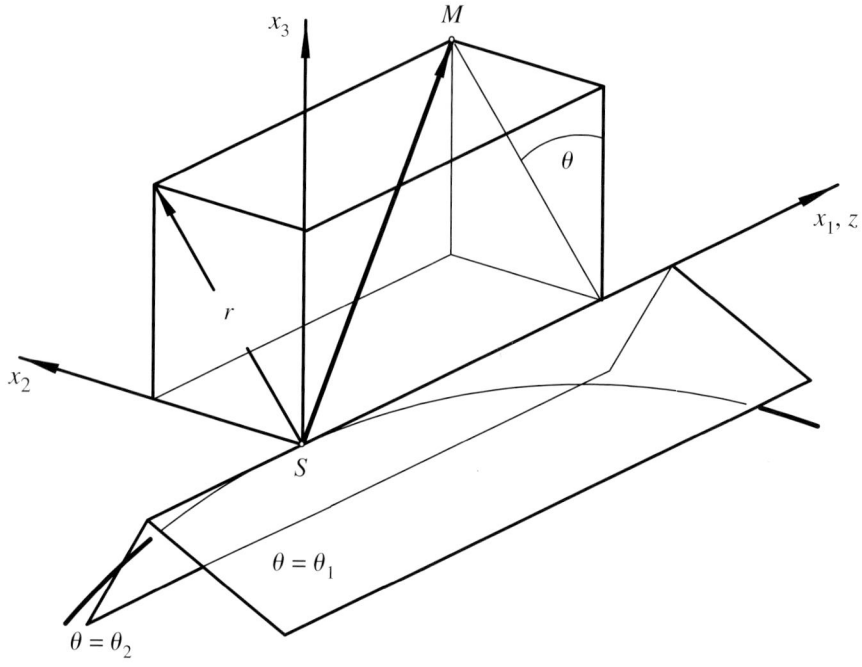

Figure 1. Cylindrical coordinates in the vicinity of point S of a smooth edge.

$$x_1 = z, \quad x_2 = r \sin \theta, \quad x_3 = r \cos \theta,$$

and the wave-vector components 137 of Chapter 1 can be written as

$$k_{1m} = k_m \cos q_{1m}, \quad k_{2m} = k_{rm} \sin \alpha_m,$$
$$k_{3m} = k_{rm} \cos \alpha_m, \quad k_{rm} = k_m \sin q_{1m}, \tag{2}$$

where k_{rm} is the projection of $\mathbf{k}_m$ on the plane $z = $ constant and α_m is the angular cylindrical coordinate of this projection (Figure 2). By substituting these equations into 136 of Chapter 1, we obtain

$$F_m = K_m \exp\{ik_m[z\cos q_{1m} + r \sin q_{1m} \cos (\theta - \alpha_m)]\}. \tag{3}$$

In contrast to the case of reflection and transmission at a regular point of interface, we cannot use equation 3 to find the unknown quantities K_m directly from interface conditions 1. However, by substituting

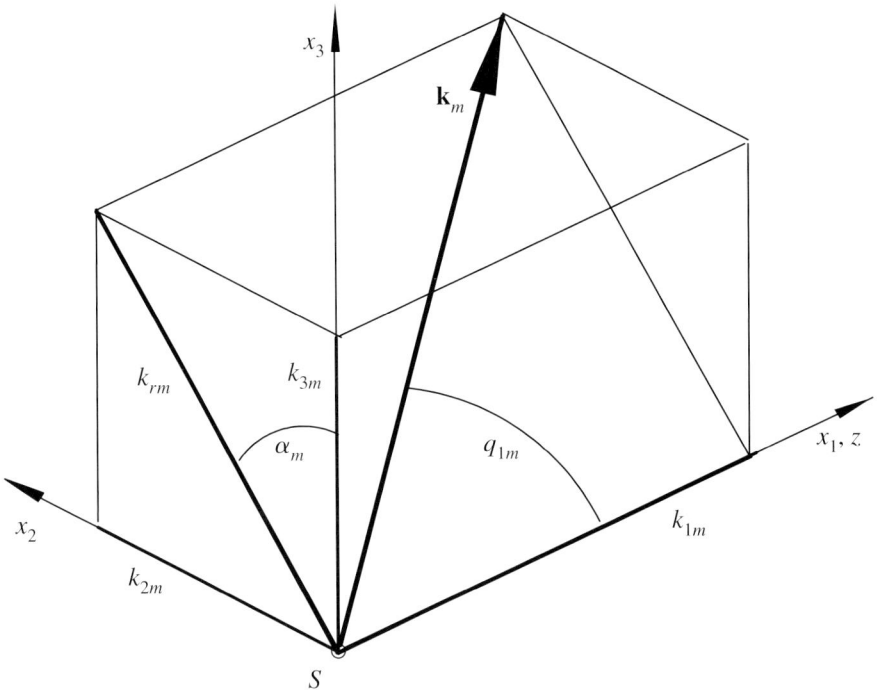

Figure 2. Wave-vector projections on the coordinate axes in the problem of edge diffraction.

equation 3 into 1 using condition 146 of Chapter 1 and the additional condition

$$\frac{\sin q_{1m} \cos\left[\pi/2 - (\theta_\ell - \alpha_m)\right]}{c_m} =$$
$$\frac{\sin q_{10} \cos\left[\pi/2 - (\theta_\ell - \alpha_0)\right]}{c_0}$$
$$\text{with} \quad \ell = 1, 2, \tag{4}$$

we can obtain the algebraic equations for K_m with coefficients independent of the coordinates. However, they are overdetermined — four equations for two unknown quantities.

To extend reflection/transmission laws to the neighborhood of the edge, we look for a solution of equation 139 of Chapter 1 in the form of the superposition of plane waves 3:

$$F_m = f_m(r, \theta)\exp(ik_m z \cos q_{1m}), \tag{5}$$

$$f_m(r, \theta) = \int_\Gamma \Phi_m(\alpha) \exp[ik_m r \sin q_{1m} \cos(\theta - \alpha)] d\alpha, \tag{6}$$

where Γ is some contour allowing the existence of the integral.

By substituting equation 5 above into equation 139 of Chapter 1, we obtain the 2D Helmholtz equations

$$\frac{\partial^2 f_m}{\partial r^2} + \frac{1}{r} \frac{\partial f_m}{\partial r} + \frac{1}{r^2} \frac{\partial^2 f_m}{\partial \theta^2} + k_{rm}^2 f_m = 0, \tag{7}$$

$$k_{rm} = k_m \sin q_{1m}. \tag{8}$$

After substituting equation 5 into 1 and using condition 146 of Chapter 1, we obtain the 2D boundary conditions

$$f_0 + f_1 = f_2, \quad \frac{1}{\rho_1} \frac{\partial}{\partial \theta} (f_0 + f_1) = \frac{1}{\rho_2} \frac{\partial f_2}{\partial \theta} \quad \text{with} \quad \theta = \theta_\ell \quad (\ell = 1, 2). \tag{9}$$

Thus, the reflection/transmission problem in the neighborhood of the edge reduces to integration of equations 7 under conditions 9. Representation of the required fields in the form 5 reduces the original 3D problem to a 2D problem in the polar coordinate space (r, θ). Equations 7 through 9 describe the wave motion in the plane $z =$ constant. The true wave velocities c_m are replaced by apparent velocities, $c_m / \sin q_{1m}$, i.e., by velocities of the wavefront intersections in the plane $z =$ constant. Because the required solutions are represented by superposition of plane waves 6, equations 7 are satisfied automatically. Relationships 4 correspond to Snell's law in 2D space (r, θ) with apparent velocities $c_m / \sin q_{1m}$.

Diffracted rays

Equation 146 of Chapter 1 is the only limitation on the directions of reflected/transmitted rays at the edge. Condition 4 above imposes no limitations. All rays complying with equation 146 of Chapter 1 are allowable. These are known as diffracted rays.

This can be explained as follows: As opposed to the case of reflection/transmission at a regular interface point, we cannot determine the tangent plane at the edge unambiguously. Indeed, we can make line L coincide with the tangent to the edge because this line is a common element of both half planes approximating the interface. However, there is no other line L' that

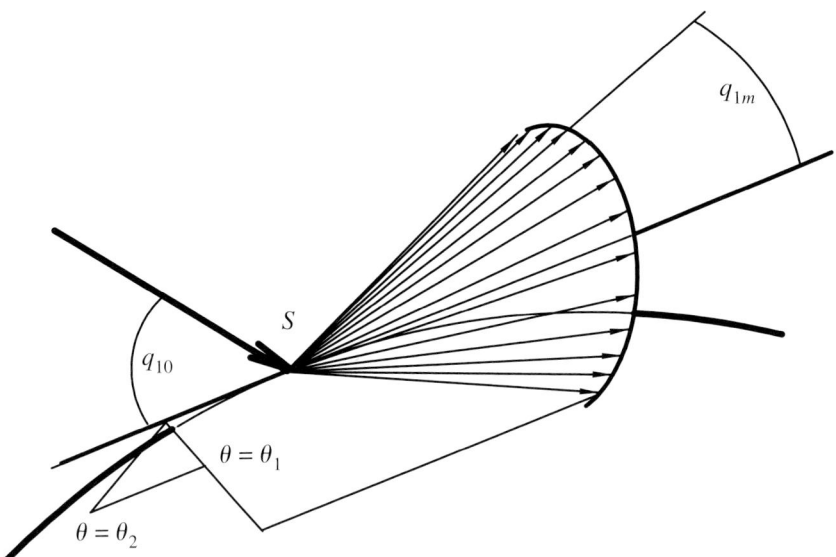

Figure 3. Cone of diffracted rays in a neighborhood of the point of incidence (Klem-Musatov, 1994).

could be a common element of the same half planes. That is why there are no other limitations on the directions of reflected/transmitted rays.

This fact is known as the *law of edge diffraction* (Keller, 1962), which reads as follows: Let an incident ray make an acute angle q_{10} with the tangent to the edge. The set of secondary rays forms a cone with its vertex at the point of incidence. Its apex angle is $2q_{1m}$, in which q_{1m} and q_{10} are connected by equation 146 of Chapter 1. The incident ray and the cone lie on opposite sides of the plane normal to the edge (Figure 3).

This law holds true within a small neighborhood of the point of incidence, where it is possible to neglect the curvature of the rays.

Diffracted rays can be continued outside the neighborhood of the edge, according to equation 17 of Chapter 1. A set of such rays forms a congruence with parameters (θ, q_{1m}) in such a way that every pair of fixed values $\theta(S) = $ constant and $q_{1m}(S) = $ constant specifies an individual ray emitted from point S of the edge.

Edge wave

To reduce the reflection/transmission problem to equations 7 through 9, we must look for a solution in the form of waves propagating along diffracted rays. It is easy to show that the existence of such waves follows from equations 5 and 6 when $\omega \to \infty$.

Equation 6, which depends on parameter ω as $\omega \to \infty$, contains the integral of a rapidly oscillating function. The asymptotic value of such an integral can be described by the well-known formula of the saddle-point method (Felsen and Marcuvitz, 1973; Bleistein and Handelsman, 1975):

$$\int_{\Gamma} \Phi(\alpha) \exp[i\rho \cos(\alpha - \theta)] d\alpha \sim \sqrt{\frac{2\pi}{\rho}} \Phi(\theta) \exp[i(\rho + \psi)] \quad \text{for} \quad \rho \to \infty,$$

$$(10)$$

where ψ depends on the form of contour Γ.

Using this approximation for integral 6, we obtain the asymptotic value of expression 5:

$$F_m = \Phi'_m \exp(i\omega\tau_m), \quad \Phi'_m = \chi'_m \sqrt{\frac{c_m}{J_m}}$$

with the condition that $k_m r \sin q_{1m} \to \infty$, $\quad$ (11)

$$\chi'_m = \sqrt{\frac{2\pi}{k_m}} \Phi_m(\theta) \exp(i\psi), \quad (12)$$

$$J_m/c_m = r \sin q_{1m}, \quad \tau_m = (r \sin q_{1m} + z \cos q_{1m})/c_m. \quad (13)$$

It is easy to see that quantity τ_m here corresponds to the eikonal of a wave propagating along diffracted rays. Quantity Φ'_m satisfies the transport equation of the type described by equation 19 of Chapter 1, i.e., it corresponds to the amplitude of a propagating wave. Thus the reflection/transmission phenomenon at the edge yields a new type of waves propagating along diffracted rays. Such waves are called edge-diffracted waves.

The geometric theory of diffraction

The edge-wave concept appears in the elementary analysis of the reflection/transmission problem for plane waves at a wedge-shaped interface. This concept can be expanded easily to a more general case. Indeed, equation 11 holds true outside the neighborhood of the edge, if functions Φ'_m and τ_m are continued there as solutions of equations 19 and 7 of Chapter 1, respectively.

The systematic description of edge effects as waves emitted by edges is introduced by the geometric theory of diffraction (Keller, 1962). This theory is designed to find the asymptotic expansions of diffracted waves into ray series with $\omega \to \infty$. However, we limit ourselves to consideration of the high-frequency asymptotic description of edge waves of relative order $O(\omega^{-1})$. The geometric theory of diffraction describes such an approximation as follows:

A stationary edge wave is determined according to equations

$$F'_m = \Phi'_m \exp(i\omega\tau_m), \quad \nabla\tau_m = \mathbf{e}_m / c_m, \tag{14}$$

in which $\mathbf{e}_m$ is the unit vector of the tangent to the diffracted ray. The congruence of diffracted rays can be obtained by using the law of edge diffraction. Therefore, the problem of mathematical description of the edge waves can be reduced to finding their amplitudes.

Let the edge wave at an arbitrary point M of some diffracted ray be determined by equation 14, where Φ'_m satisfies the transport equation

$$2\nabla\tau_m \cdot \nabla\Phi'_m + \Phi'_m\Delta\tau_m = 0. \tag{15}$$

Let S be a point of the same diffracted ray in the near vicinity of the edge. Solutions of this equation at points M and S are connected by the relationship

$$\Phi'_m(M) = \frac{\Phi'_m(S)}{L_m}. \tag{16}$$

However, the quantity $\Phi'_m(S)$ in the near vicinity of the edge can be described by the second expression of 11, where χ'_m and J_m are determined by equations 12 and 13. One can see that it is necessary to know function $\Phi_m(\theta)$ and parameter ψ to find amplitude 16. To determine these quantities, it is necessary to obtain a solution of equations 7 through 9 in the form 6.

Thus, the edge wave (within the framework of the geometric theory of diffraction) can be found by its continuation from the neighborhood of the edge in accordance with ray-theory expression 16. Its amplitude in the neighborhood of the edge must be obtained from the solution of some canonical problem of the mathematical theory of diffraction. This solution must approximate the edge wave in the edge neighborhood locally. The simplest canonical problem is diffraction of the plane wave from a system of wedge-shaped regions with a common edge.

The geometric theory of diffraction is valid in regions of space where the edge-wave amplitude satisfies transport equation 15. However, the latter fails in shadow-boundary neighborhoods of reflected/transmitted waves. When approaching a shadow boundary, function $\Phi_m(\theta)$ in equation 12 tends to infinity, and the second expression in 11 makes no sense. Therefore, expression 16 fails. Domains in which expression 16 fails are called boundary layers.

Equation 12 shows that the edge-wave amplitude is $\sqrt{\omega}$ times less than the ordinary reflected/transmitted-wave amplitude. Therefore, it is possible to neglect the edge wave compared with the reflected/transmitted wave when $\omega \rightarrow \infty$. However, as we shall see later, the edge-wave amplitude in the boundary layer is comparable with the reflected/transmitted-wave amplitude when $\omega \rightarrow \infty$. Therefore, consideration of edge waves in boundary layers is important for applications.

The equation of the boundary layer

Neighborhood of the shadow boundary

In this section, we will study the edge-diffraction phenomenon in the close neighborhood of the shadow boundary of an individual wave. We will consider a high-frequency solution ($\omega \rightarrow \infty$) of the stationary wave equation

$$\left(\Delta + k^2\right)f = 0, \quad k = \omega/c, \tag{17}$$

in which the wave velocity $c(M)$ depends on point M of an inhomogeneous medium.

To find a solution of this equation within the framework of ray theory, it is necessary to introduce a congruence of rays. Let such a congruence be given by the two-parameter set (α_0, β_0) so that each fixed pair of values $\alpha_0 = $ constant, $\beta_0 = $ constant specifies an individual ray. Let τ_0 be the eikonal along this congruence. Then equation 17 can be reduced to a transport equation that describes the amplitude of a wave propagating along the congruence (α_0, β_0).

Suppose such a wave has a shadow boundary caused by the interruption of regular propagation or reflection/transmission at an interface containing an edge. Then in conformity with the law of edge diffraction, there must be a congruence of diffracted rays spreading from the edge and a corresponding edge wave at this congruence. Because the total field of these two waves must satisfy equation 17, we can try to find the total field by direct integration of this equation. Such a problem can be solved if we limit ourselves by considering equation 17 within the close neighborhood of the shadow boundary. To do this, we must introduce the concept of the neighborhood of the shadow boundary.

Let us begin with the formal introduction of a congruence of diffracted rays. The expression $\alpha_0 = $ constant gives a surface formed by the rays of the congruence (α_0, β_0) (Figure 4). To simplify the geometric interpretation,

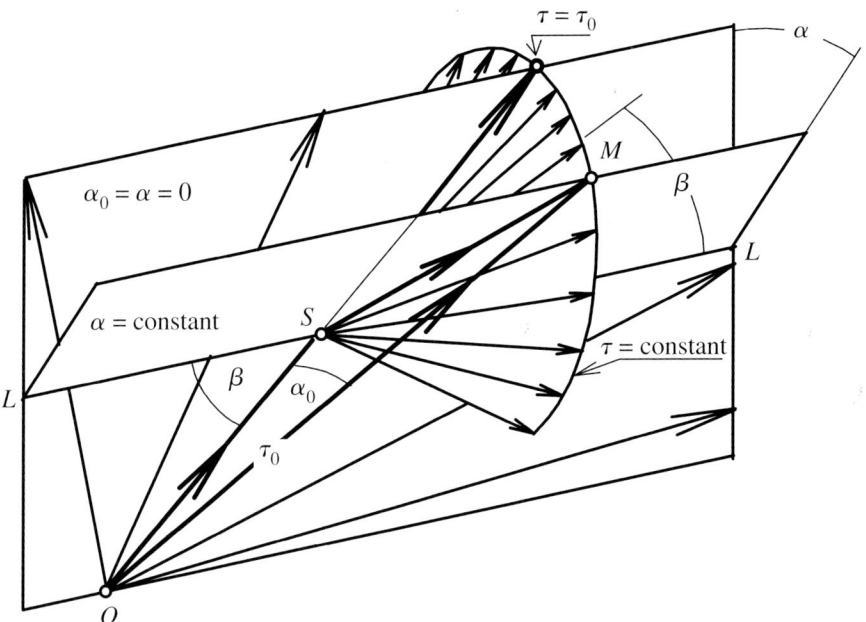

Figure 4. Definition of the neighborhood of a shadow boundary ($\alpha_0 = 0$) connected with the edge L. OM is a ray emitted by the source O to a current point M in space, and SM is a diffracted ray.

we use rectilinear lines in this figure. However, all of the following considerations hold true for the curves: Let L be an infinite smooth line at a particular surface α_0 = constant that we choose as $\alpha_0 = 0$. Such a line can be called a guide of the surface $\alpha_0 = 0$. Let L cross each ray of this surface only once. Let points O and S belong to a ray on this surface. Let $\tau_0(S) > \tau_0(O)$ and require that point S be constrained to lie on line L, i.e., S corresponds to the point of incidence of the ray OS at the line L.

Let the space line SM (where M is a current point of space) satisfy the ray equation 17 of Chapter 1. Let rays OS and SM lie on different sides of a plane normal to the line L at point S. Choose ray OS so that for a given M, the tangents to rays OS and SM form equal angles with the tangent to L at S. One can see that the directions of rays OS and SM are connected by the law of diffraction. Then line L corresponds to the edge, line OS to the incident ray, and SM to the diffracted ray.

By changing the position of point M in space, we can construct a congruence of rays SM connected with the line L. We can describe such a congruence by the two-parameter set (α, β), in which each fixed pair of values $\alpha = $ constant, $\beta = $ constant singles out an individual ray. Either of the expressions $\alpha = $ constant or $\beta = $ constant describes a surface formed by the rays of this congruence. We can choose parameters so that any surface $\alpha = $ constant is semi-infinite and contains the line L as an edge. We set $\alpha = 0$ on that surface which coincides with the surface $\alpha_0 = 0$. We can choose the parameter β so that the surface $\beta = $ constant corresponds to the cone of rays emitted by point S in agreement with the law of diffraction (Figure 4).

Now we can introduce the concept of the shadow-boundary neighborhood. We have introduced two congruences of rays (α_0, β_0) and (α, β) with the common surface $\alpha = 0$. At this surface, the following relationship holds true:

$$\tau_0 = \tau, \tag{18}$$

in which τ_0 and τ are the eikonals along the congruences (α_0, β_0) and (α, β), respectively. This expression can be considered as an implicit equation of the surface. One can identify the above-described congruences with the families of rays taking part in the edge-diffraction description and consider (α_0, β_0) and (α, β) as the congruences of reflected/transmitted and diffracted rays, respectively. One can consider surface 18 as the shadow boundary.

One can try to find high-frequency asymptotic solutions of equation 17 as $\omega \rightarrow \infty$, connected with each above-mentioned congruences separately, by using ray theory or the geometric theory of diffraction. However, such approaches fail within the near vicinity of the common surface of both congruences. The proximity of an arbitrary point M to the surface 18 can be estimated by the quantity

$$T = \left| \omega \left[\tau_0(M) - \tau(M) \right] \right|. \tag{19}$$

We call a region the boundary layer or the neighborhood of the shadow boundary if its points satisfy the following inequality:

$$\left| \omega \left[\tau_0(M) - \tau(M) \right] \right| < C \quad \text{as} \quad \omega \rightarrow \infty, \tag{20}$$

where constant C does not depend on ω. We will look for solutions of equation 17 in region 20. The problem under consideration consists of integrating equation 17 in the special curvilinear coordinate system (τ, α, β) connected with the congruence of diffracted rays. Condition 20 will be used for local approximation of the resulting equations. The existence of the initial congruence (α_0, β_0) can be taken into account in the initial form of the desired solution.

The high-frequency approximation

Look for a solution of equation 17 of the form

$$f = AG_0 \exp(i\omega\tau), \quad (\nabla\tau)^2 = \frac{1}{c^2}, \tag{21}$$

in which G_0 satisfies the transport equation at the congruence (α_0, β_0), i.e.,

$$2\nabla\tau_0 \cdot \nabla G_0 + G_0\Delta\tau_0 = 0. \tag{22}$$

Substituting equations 21 and 22 into 17, we obtain the equation for the unknown function A

$$G_0\Delta A + B_1\nabla A + B_2 A = 0, \tag{23}$$

where

$$B_1 = 2(i\omega G_0\nabla\tau + \nabla G_0), \tag{24}$$

and

$$B_2 = i\omega(G_0\Delta\tau + 2\nabla\tau \cdot \nabla G_0) + \Delta G_0. \tag{25}$$

Rewrite coefficient B_2 in a more convenient form. To do this, take the scalar products

$$\nabla\tau_0 \cdot \nabla G_0 = \frac{|\nabla G_0|\cos(\nabla\tau_0, \nabla G_0)}{c}, \quad \nabla\tau \cdot \nabla G_0 = \frac{|\nabla G_0|\cos(\nabla\tau, \nabla G_0)}{c} \tag{26}$$

and divide them term by term

$$\nabla\tau \cdot \nabla G_0 = \mu\nabla\tau_0 \cdot \nabla G_0, \tag{27}$$

where

$$\mu = \frac{\cos(\nabla\tau, \nabla G_0)}{\cos(\nabla\tau_0, \nabla G_0)}. \tag{28}$$

Then solve equation 22 for $\nabla\tau_0 \cdot \nabla G_0$:

$$\nabla\tau_0 \cdot \nabla G_0 = -\frac{G_0\Delta\tau_0}{2} \tag{29}$$

and substitute this expression into equation 27. Then

$$\nabla\tau \cdot \nabla G_0 = -\frac{\mu G_0 \Delta\tau_0}{2}. \tag{30}$$

By substituting this expression into equation 25, we obtain

$$B_2 = i\omega G_0(\Delta\tau - \mu\Delta\tau_0) + \Delta G_0. \tag{31}$$

Because we consider the case $\omega \to \infty$, we can neglect the small items in equations 24 and 31. On the strength of the inequalities

$$|\nabla\tau| \gg \left|\frac{\nabla G_0}{\omega G_0}\right|, \quad |\Delta\tau - \mu\Delta\tau_0| \gg \left|\frac{\Delta G_0}{\omega G_0}\right| \quad \text{as} \quad \omega \to \infty, \tag{32}$$

we can use the approximate expressions

$$B_1 = 2i\omega G_0 \nabla\tau, \quad B_2 = i\omega G_0(\Delta\tau - \mu\Delta\tau_0) \quad \text{as} \quad \omega \to \infty. \tag{33}$$

By substituting these expressions into equation 23, we obtain the differential equation

$$\Delta A + 2i\omega\nabla\tau \cdot \nabla A + i\omega(\Delta\tau - \mu\Delta\tau_0)A = 0. \tag{34}$$

We will express the differential operations in coordinate form to simplify this equation.

Ray coordinates

Use ray coordinates (τ, α, β). Because this coordinate system is orthogonal, the following well-known expressions for the differential operations hold true:

$$\nabla\varphi = \nabla_\tau\varphi + \nabla_\alpha\varphi + \nabla_\beta\varphi, \tag{35}$$

$$\Delta\varphi = \Delta_\tau\varphi + \Delta_\alpha\varphi + \Delta_\beta\varphi, \tag{36}$$

$$\nabla_\gamma\varphi = \frac{1}{\sqrt{g_{\gamma\gamma}}} \frac{\partial\varphi}{\partial\gamma} \mathbf{i}_\gamma, \tag{37}$$

$$\Delta_\gamma\varphi = \frac{1}{J} \frac{\partial}{\partial\gamma}\left(\frac{J}{g_{\gamma\gamma}} \frac{\partial\varphi}{\partial\gamma}\right), \tag{38}$$

$$g_{\gamma\gamma} = \sum_{n=1}^{3}\left(\frac{\partial x_n}{\partial \gamma}\right)^2, \tag{39}$$

$$J = D(x_1, x_2, x_3)/D(\tau, \alpha, \beta), \tag{40}$$

where $\mathbf{i}_\gamma$ are the unit vectors of the Cartesian coordinate system (x_1, x_2, x_3).

Let us mention further a useful relationship that holds true in ray coordinates. Take $\varphi = \tau$ in equation 37. Then

$$\nabla\tau = \nabla_\tau\tau = \frac{\mathbf{i}_\tau}{\sqrt{g_{\tau\tau}}}. \tag{41}$$

Take the scalar products $\nabla\tau_0 \cdot \nabla G_0$ and $\nabla\tau \cdot \nabla G_0$, defined as

$$\nabla\tau = \left(\nabla_\tau + \nabla_\alpha + \nabla_\beta\right)\tau, \quad \nabla\varphi = \left(\nabla_\tau + \nabla_\alpha + \nabla_\beta\right)\varphi, \tag{42}$$

where φ is an arbitrary scalar field. Using equations 37 and 41 above and equation 53 of Chapter 1 and taking into consideration the relationships

$$\mathbf{i}_\tau \cdot \mathbf{i}_\alpha = 0, \quad \mathbf{i}_\tau \cdot \mathbf{i}_\beta = 0, \quad \mathbf{i}_\tau \cdot \mathbf{i}_\tau = 1, \tag{43}$$

we obtain the relationship

$$\nabla\tau \cdot \nabla\varphi = \nabla_\tau\tau \cdot \left(\nabla_\tau + \nabla_\alpha + \nabla_\beta\right)\varphi = \nabla_\tau\tau \cdot \nabla_\tau\varphi = \frac{1}{c^2}\frac{\partial\varphi}{\partial\tau}. \tag{44}$$

The boundary layer in ray coordinates

Rewrite equation 20 in ray coordinates. To do this, consider the eikonal τ_0 as a function of the ray coordinates $\tau_0 = \tau_0(\tau, \alpha, \beta)$ and expand it in a power series with respect to the coordinate α near the surface $\alpha = 0$,

$$\tau_0(\tau, \alpha, \beta) = \sum_{k=0}^{\infty} a_k(\tau, 0, \beta)\alpha^k, \tag{45}$$

$$a_k(\tau, 0, \beta) = \frac{1}{k!}\left(\frac{\partial^k \tau_0}{\partial \alpha^k}\right)_{\alpha=0}. \tag{46}$$

The first coefficient of this series satisfies the obvious condition

$$a_0(\tau, 0, \beta) = \tau_0(\tau, 0, \beta) = \tau, \tag{47}$$

following from equation 18. The second coefficient is zero, in accordance with equation 46,

$$a_1(\tau, 0, \beta) = \left(\frac{\partial \tau_0}{\partial \alpha}\right)_{\alpha=0} = \left[|\nabla \tau_0| \cos(\nabla \tau_0, \mathbf{i}_\alpha)\right]_{\alpha=0}$$

$$= \frac{\cos(\mathbf{i}_\tau, \mathbf{i}_\alpha)}{c} = 0, \tag{48}$$

because unit vectors $\mathbf{i}_\tau$ and $\mathbf{i}_\alpha$ are mutually orthogonal.

Using equations 47 and 48, rewrite equation 45 as

$$\tau_0(M) - \tau(M) = \sum_{k=2}^{\infty} a_k(\tau, 0, \beta)\alpha^k. \tag{49}$$

Suppose that coefficient $a_2(\tau, 0, \beta)$ does not equal zero for the values of τ and β being considered. Then condition 20 with $\omega \to \infty$ can be satisfied only when $\alpha \to 0$. Hence in equation 49, we can neglect all items with $k > 2$. Then

$$\tau_0(M) - \tau(M) = a_2(\tau, 0, \beta)\alpha^2. \tag{50}$$

By substituting this expression into condition 20, we obtain the definition of the boundary layer in ray coordinates,

$$\left|\omega\alpha^2 a_2(\tau, 0, \beta)\right| < C \quad \text{as} \quad \omega \to \infty. \tag{51}$$

Look for a solution of equation 34 in the boundary layer determined by condition 51. Because $\omega \to \infty$, this condition can be satisfied only for $\alpha \to 0$, and we essentially can simplify the differential equation under consideration. To do this, we will use the new nonorthogonal coordinate system (z, α, β) in which the quantity

$$z = i\omega(\tau_0 - \tau), \quad \tau_0 = \tau_0(\tau, \alpha, \beta) \tag{52}$$

is a measure of the proximity to surface 18 and look for a solution of equation 34 in the form

$$A = A(z, \alpha, \beta). \tag{53}$$

The operations ∇A and ΔA

Express the differential operations ∇A and ΔA in coordinates (z, α, β). By differentiating equation 53 as a composite function, we obtain

$$\nabla A = \sum_{\gamma=z,\alpha,\beta} A_\gamma \nabla\gamma,$$

$$\Delta A = \nabla\cdot(\nabla A) = \sum_{\gamma=z,\alpha,\beta} \nabla\cdot(A_\gamma \nabla\gamma) = \sum_{\gamma=z,\alpha,\beta} \left(\nabla A_\gamma \cdot \nabla\gamma + A_\gamma \Delta\gamma\right)$$

$$= \sum_{\gamma=z,\alpha,\beta} \left[\sum_{\mu=z,\alpha,\beta} (A_{\gamma\mu}\nabla\gamma\cdot\nabla\mu + A_\gamma\Delta\gamma) \right], \tag{54}$$

where subscript indices denote derivatives with respect to corresponding variables.

Because ray coordinates are orthogonal, we have

$$\nabla\gamma\cdot\nabla\mu = 0 \quad \text{with} \quad \gamma \neq \mu \quad (\gamma, \mu = \tau, \alpha, \beta). \tag{55}$$

Taking into consideration equation 55 and relationship $A_{\gamma\mu} = A_{\mu\gamma}$, we obtain

$$\Delta A = (\nabla z)^2 A_{zz} + (\nabla\alpha)^2 A_{\alpha\alpha} + (\nabla\beta)^2 A_{\beta\beta} + 2\nabla z\cdot\nabla\alpha A_{z\alpha}$$

$$+ 2\nabla z\cdot\nabla\beta A_{z\beta} + \Delta z A_z + \Delta\alpha A_\alpha + \Delta\beta A_\beta. \tag{56}$$

To find the coefficients at the derivatives, write the gradient and Laplacian of the function $z(\tau, \alpha, \beta)$ in orthogonal ray coordinates

$$\nabla z = \sum_{\gamma=\tau,\alpha,\beta} z_\gamma \nabla\gamma,$$

$$\Delta z = \nabla\cdot(\nabla z) = \sum_{\gamma=\tau,\alpha,\beta} \nabla\cdot(z_\gamma \nabla\gamma)$$

$$= \sum_{\gamma=\tau,\alpha,\beta} \left(\sum_{\mu=\tau,\alpha,\beta} z_{\gamma\mu}\nabla\gamma\cdot\nabla\mu + z_\gamma\Delta\gamma \right). \tag{57}$$

Taking into consideration equation 55, we obtain

$$\Delta z = z_{\tau\tau}(\nabla\tau)^2 + z_{\alpha\alpha}(\nabla\alpha)^2 + z_{\beta\beta}(\nabla\beta)^2 + z_\tau\Delta\tau + z_\alpha\Delta\alpha + z_\beta\Delta\beta. \tag{58}$$

By using equations 50 and 52, quantity z can be represented as

$$z = \alpha^2 p(\tau, \beta), \quad p = i\omega a_2(\tau, 0, \beta). \tag{59}$$

Differentiating this expression, we have

$$z_\alpha = \frac{2z}{\alpha}, \quad z_{\alpha\alpha} = \frac{2z}{\alpha^2}, \quad z_\gamma = p_\gamma \alpha^2, \quad z_{\gamma\gamma} = p_{\gamma\gamma} \alpha^2 \quad \text{with} \quad \gamma = \tau, \beta.$$

(60)

Using equations 55 and 57, we obtain

$$(\nabla z)^2 = \nabla z \cdot \nabla z = z_\tau^2 (\nabla \tau)^2 + z_\alpha^2 (\nabla \alpha)^2 + z_\beta^2 (\nabla \beta)^2.$$

By substituting equations 60 into this expression, we obtain

$$(\nabla z)^2 = \frac{1}{\alpha^2} \left[4z^2 (\nabla \alpha)^2 + \alpha^6 \sum_{\gamma = \tau, \beta} p_\gamma^2 (\nabla \gamma)^2 \right].$$

(61)

Using equations 55 and 57, with $\gamma \neq \alpha$, we obtain

$$\nabla z \cdot \nabla \gamma = (z_\tau \nabla \tau + z_\alpha \nabla \alpha + z_\beta \nabla \beta) \cdot \nabla \gamma = z_\tau \nabla \tau \cdot \nabla \gamma + z_\beta \nabla \beta \cdot \nabla \gamma$$

because $\nabla \alpha \cdot \nabla \gamma = 0$ with $\gamma \neq \alpha$. By substituting equations 60 into this expression, we obtain

$$\nabla z \cdot \nabla \gamma = \alpha^2 (p_\tau \nabla \tau + p_\beta \nabla \beta) \cdot \nabla \gamma = \alpha^2 p_\gamma (\nabla \gamma)^2 \quad \text{for} \quad \gamma = \tau, \beta$$

(62)

because $\nabla \tau \cdot \nabla \gamma = 0$ for $\gamma = \beta$, and $\nabla \beta \cdot \nabla \gamma = 0$ for $\gamma = \tau$.

By substituting equations 60 into equation 58, we obtain

$$\Delta z = \frac{1}{\alpha^2} \left\{ 2z (\nabla \alpha)^2 + 2z\alpha \Delta \alpha + \alpha^2 \sum_{\gamma = \tau, \beta} \left[p_{\gamma\gamma} (\nabla \gamma)^2 + p_\gamma \Delta \gamma \right] \right\}.$$

(63)

The scalar product $\nabla \tau \cdot \nabla A$

Express quantity $\nabla \tau \cdot \nabla A$ from equation 34 in the coordinates (z, α, β). By using equations 54 and 55, we obtain

$$\nabla \tau \cdot \nabla A = \nabla \tau \cdot (A_z \nabla z + A_\alpha \nabla \alpha + A_\beta \nabla \beta) = A_z \nabla \tau \cdot \nabla z.$$

(64)

Transform the scalar product $\nabla \tau \cdot \nabla z$ to a more convenient form by using the representation

$$2\nabla\tau\cdot\nabla z = (2\nabla\tau - \nabla\tau_0 + \nabla\tau_0)\cdot\nabla z = \left[\nabla(\tau - \tau_0) + \nabla(\tau + \tau_0)\right]\cdot\nabla z$$
$$= \nabla(\tau - \tau_0)\cdot\nabla z + \nabla(\tau + \tau_0)\cdot\nabla z. \tag{65}$$

From equation 52, it follows that

$$\nabla(\tau - \tau_0) = -\frac{\nabla z}{i\omega}. \tag{66}$$

However, by performing the scalar product formally, we obtain

$$\nabla(\tau + \tau_0)\cdot\nabla z = -i\omega\nabla(\tau + \tau_0)\cdot\nabla(\tau - \tau_0)$$
$$= -i\omega(\nabla\tau + \nabla\tau_0)\cdot(\nabla\tau - \nabla\tau_0) = -i\omega\left[(\nabla\tau)^2 - (\nabla\tau_0)^2\right] = 0 \tag{67}$$

because $(\nabla\tau)^2 = (\nabla\tau_0)^2 = 1/c^2$, where c is the propagation velocity.
Substituting equations 66 and 67 into 65, we obtain

$$\nabla\tau\cdot\nabla z = -\frac{(\nabla z)^2}{2i\omega}. \tag{68}$$

Substituting this expression into equation 64, we have

$$\nabla\tau\cdot\nabla A = -\frac{A_z(\nabla z)^2}{2i\omega}. \tag{69}$$

The equation of the boundary layer

Substitute expressions 56 and 69 with coefficients defined by equations 61 through 63 into 34 and write the result as

$$b_1 A_{zz} + b_2 A_{\alpha\alpha} + b_3 A_{\beta\beta} + b_4 A_{z\alpha} + b_5 A_{z\beta} + b_6 A_z$$
$$+ b_7 A_\alpha + b_8 A_\beta + b_9 A = 0, \tag{70}$$

in which

$$b_1 = 4z^2(\nabla\alpha)^2 + \alpha^6 \sum_{\gamma=\tau,\beta} p_\gamma^2(\nabla\gamma)^2,$$

$$b_2 = \alpha^2(\nabla\alpha)^2,$$

$$b_3 = \alpha^2(\nabla\beta)^2,$$

$$b_4 = 2\alpha^4 p_\alpha(\nabla\alpha)^2,$$

$$b_5 = 2\alpha^4 p_\beta (\nabla \beta)^2,$$

$$b_6 = 2z(1 - 2z)(\nabla \alpha)^2 + 2\alpha z \Delta \alpha$$

$$+ \alpha^2 \sum_{\gamma = \tau, \beta} \left[p_{\gamma\gamma}(\nabla \gamma)^2 + p_\gamma \Delta \gamma + \alpha^4 p_\gamma^2 (\nabla \gamma)^2 \right],$$

$$b_7 = \alpha^2 \Delta \alpha,$$

$$b_8 = \alpha^2 \Delta \beta,$$

$$b_9 = i\omega\alpha^2 (\Delta \tau - \mu \Delta \tau_0).$$

The coefficients of this equation are functions of 3D space (z, α, β). In the vicinity of the coordinate surface $\alpha = 0$, i.e., in region 51, these coefficients are regular, slowly changing functions of α. Therefore, we can take the approximation

$$b_n(z, \alpha, \beta) = b_n(z, 0, \beta) \quad \text{with} \quad n = 1, 2, \ldots 9. \tag{71}$$

Then

$$b_1 = 4z^2(\nabla \alpha)^2, \quad b_6 = 2z(1 - 2z)(\nabla \alpha)^2,$$

$$b_2 = b_3 = b_4 = b_5 = b_7 = b_8 = 0. \tag{72}$$

Consider the coefficient b_9. When $\alpha \to 0$, we have in equation 28

$$\mu = \left[\mu(\tau, \alpha, \beta) \right]_{\alpha=0} = \mu(\tau, 0, \beta) = 1 \tag{73}$$

and

$$b_9 = i\omega\alpha^2 (\Delta \tau - \Delta \tau_0) = \alpha^2 \Delta \left[i\omega(\tau - \tau_0) \right].$$

Taking into account equation 52, rewrite this expression as $b_9 = -\alpha^2 \Delta z$. By substituting equation 63 into this expression and taking $\alpha = 0$, we obtain

$$b_9 = -2z(\nabla \alpha)^2. \tag{74}$$

Substituting expressions 72 and 74 into 70, we obtain the ordinary differential equation

$$z A_{zz} + \left(\frac{1}{2} - z \right) A_z - \frac{1}{2} A = 0. \tag{75}$$

This is the differential equation of the boundary layer.

Linearly independent solutions

Kummer's equation

The differential equation

$$z \frac{d^2 A}{dz^2} + (\gamma - z) \frac{dA}{dz} - \delta A = 0, \tag{76}$$

in which γ and δ are arbitrary numbers, is a confluent hypergeometric equation also known as Kummer's equation (Abramowitz and Stegun, 1972). It has two linearly independent solutions,

$$A = \Phi(\delta, \gamma; z), \tag{77}$$

and

$$A = z^{1-\gamma}\Phi(\delta + \gamma, 2 - \gamma; z), \tag{78}$$

where the functions $\Phi(\delta, \gamma; z)$ are called confluent hypergeometric functions or Kummer's functions. Equation 75 is a particular case of Kummer's equation where

$$\gamma = \delta = 1/2. \tag{79}$$

In this case, functional relationships

$$\Phi\left(\frac{1}{2}, \frac{1}{2}; z\right) = \exp z,$$

$$\sqrt{z}\,\Phi\left(1, \frac{3}{2}; z\right) = \frac{\exp z}{2}\left[\sqrt{\pi} - \Gamma\left(\frac{1}{2}, z\right)\right] \tag{80}$$

apply when the incomplete gamma function

$$\Gamma\left\{\frac{1}{2}, z\right\} = \int\limits_{z}^{\infty} \frac{\exp(-t)}{\sqrt{t}}\, dt \tag{81}$$

is involved.

Using expressions 80, we can represent two linearly independent solutions of equation 75 as

$$A = \exp z, \tag{82}$$

$$A = \frac{\exp z}{2} \left[\sqrt{\pi} - \Gamma\!\left(\frac{1}{2}, z\right) \right]. \tag{83}$$

The same formulas have been obtained for two linearly independent asymptotic solutions of equations of motion of 2D anisotropic media (Druzhinin and Aizenberg, 1990) and have been generalized for the 3D scalar case by Aizenberg (1992, 1993a, 1993b).

The first solution

Substituting equation 82 into 21, we obtain

$$f = CG_0 \exp(i\omega\tau + z), \tag{84}$$

where C does not depend on z. Substituting the quantity z from equation 52 into this expression, we obtain

$$f = CG_0 \exp(i\omega\tau_0). \tag{85}$$

Because the coefficient G_0 satisfies the transport equation 22, this solution corresponds to the asymptotic ray-theory solution.

The second solution

Consider the properties of solution 83 when $\alpha \to 0$. To do this, use the well-known properties of the incomplete gamma function (Abramowitz and Stegun, 1972):

$$\Gamma\!\left(\frac{1}{2}, 0\right) = \Gamma\!\left(\frac{1}{2}\right) = \sqrt{\pi}, \tag{86}$$

$$\frac{d}{dz}\,\Gamma\!\left(\frac{1}{2}, z\right) = -\frac{\exp(-z)}{\sqrt{z}}, \tag{87}$$

where $\Gamma(1/2)$ is a gamma function. Using equation 87 for differentiation of equation 83, we obtain

$$\frac{dA}{dz} = A + \frac{1}{2\sqrt{z}}, \quad \frac{d^2A}{dz^2} = \frac{dA}{dz} - \frac{1}{4z\sqrt{z}}. \tag{88}$$

The following derivatives of solution 83 can be found by using expressions 88:

$$\frac{\partial A}{\partial \eta} = \frac{dA}{dz}\frac{\partial z}{\partial \eta}, \quad \frac{\partial^2 A}{\partial \eta^2} = \frac{d^2 A}{dz^2}\left(\frac{\partial z}{\partial \eta}\right)^2 + \frac{dA}{dz}\frac{\partial^2 z}{\partial \eta^2}, \quad \text{for} \quad \eta = \tau, \, \alpha, \, \beta \quad (89)$$

with $\alpha = 0$. Straightforward calculations yield

$$(A)_{\alpha=0} = \left(\frac{\partial A}{\partial \eta}\right)_{\alpha=0} = \left(\frac{\partial^2 A}{\partial \eta^2}\right)_{\alpha=0} = 0 \quad \text{for} \quad \eta = \tau, \, \alpha, \, \beta, \quad (90)$$

with the exception of the derivative $\partial A/\partial \alpha$.

Thus, the solution under consideration and all these derivatives are continuous at the surface $\alpha = 0$, with the exception of the derivative $\partial A/\partial \alpha$.

In the case of $\partial A/\partial \alpha$, the calculations yield

$$\frac{\partial A}{\partial \alpha} = \sqrt{i\omega a_2(\tau, \, 0, \, \beta)}\,\frac{\alpha}{\sqrt{\alpha^2}}. \quad (91)$$

Here, the value of the coefficient

$$\frac{\alpha}{\sqrt{\alpha^2}} = \frac{\alpha}{|\alpha|} = \text{sign}\,\alpha \quad (92)$$

depends on the direction of approach to the surface $\alpha = 0$:

$$\left(\frac{\alpha}{\sqrt{\alpha^2}}\right)_{\alpha=0_{\pm}} = (\text{sign}\,\alpha)_{\alpha=0_{\pm}} = \pm 1. \quad (93)$$

Because of this, the value of the quantity 91 is dependent on direction,

$$\left(\frac{\partial A}{\partial \alpha}\right)_{\alpha=0_{\pm}} = \pm\sqrt{i\omega a_2(\tau, \, 0, \, \beta)}, \quad (94)$$

and the following relationship holds true:

$$\left(\frac{\partial A}{\partial \alpha}\right)_{\alpha=0_+} - \left(\frac{\partial A}{\partial \alpha}\right)_{\alpha=0_-} = 2\sqrt{i\omega a_2(\tau, \, 0, \, \beta)}. \quad (95)$$

Equation 95 shows that the derivative 91 has a discontinuity when $\alpha = 0$. However, the solution of the wave equation combined with all of its first and second derivatives must be continuous. Therefore, we cannot use function 83 in the vicinity of surface $\alpha = 0$ to satisfy the wave equation.

The smooth solution

Consider the following expression,

$$A^* = \begin{cases} +A & \text{when} \quad \alpha > 0, \\ -A & \text{when} \quad \alpha < 0, \end{cases} \tag{96}$$

in which A is determined by equation 83. This function satisfies Kummer's equation in both domains $\alpha < 0$ and $\alpha > 0$. It has the properties

$$(A^*)_{\alpha=0} = \left(\frac{\partial A^*}{\partial \eta}\right)_{\alpha=0} = \left(\frac{\partial^2 A^*}{\partial \eta^2}\right)_{\alpha=0} = 0 \quad \text{for} \quad \eta = \tau, \alpha, \beta, \tag{97}$$

$$\left(\frac{\partial A^*}{\partial \alpha}\right)_{\alpha=0_\pm} = \sqrt{i\omega a_2(\tau, 0, \beta)}, \tag{98}$$

and

$$\left(\frac{\partial A^*}{\partial \alpha}\right)_{\alpha=0_+} - \left(\frac{\partial A^*}{\partial \alpha}\right)_{\alpha=0_-} = 0. \tag{99}$$

One can see that function 96 is continuous along with all its first and second derivatives at surface $\alpha = 0$. Therefore, the result of substituting this function into equation 21 satisfies equation 17 in the boundary layer.

Thus we take as the secondary linearly independent solution of Kummer's equation

$$A = \text{sign } \alpha \cdot \sqrt{\pi} \left[\frac{\exp z}{2} - \frac{1}{2\sqrt{\pi}} \Gamma\left(\frac{1}{2}, z\right)\exp z\right]. \tag{100}$$

The function $W(w)$

We introduce a new special function to describe diffraction phenomena. To do this, represent the quantity z as

$$z = -\frac{i\pi w^2}{2}, \quad w = \sqrt{\frac{2iz}{\pi}} \tag{101}$$

and use the notation

$$W(w) = \frac{1}{2\sqrt{\pi}} \Gamma\left(\frac{1}{2}, -\frac{i\pi w^2}{2}\right) \exp\left(-\frac{i\pi w^2}{2}\right). \tag{102}$$

Then equation 100 can be written as

$$A = \text{sign } \alpha \cdot \sqrt{\pi} \left[\frac{1}{2}\exp\left(-\frac{i\pi w^2}{2}\right) - W(w)\right]. \tag{103}$$

From equations 52 and 101, we obtain

$$-\frac{i\pi w^2}{2} = i\omega(\tau_0 - \tau). \tag{104}$$

Substituting this expression into equation 103, we obtain

$$A = \text{sign } \alpha \cdot \sqrt{\pi} \exp(-i\omega\tau)\left[\frac{1}{2}\exp(i\omega\tau_0) - W(w)\exp(i\omega\tau)\right]. \tag{105}$$

Function W(w) possesses the properties

$$W(w) \sim W(0) + \frac{w}{\sqrt{2}}\exp\left(\frac{3\pi i}{4}\right) + O(w^2) \quad \text{as} \quad |w| \to 0, \tag{106}$$

$$W(0) = \frac{1}{2}, \tag{107}$$

$$W(w) \sim \frac{1}{\pi w \sqrt{2}}\exp\left(\frac{i\pi}{4}\right) + O\left(\frac{1}{w^2}\right) \quad \text{as} \quad |w| \to \infty, \tag{108}$$

$$W(i|w|) = \overline{W(|w|)}, \tag{109}$$

$$\frac{dW(w)}{dw} = -\frac{1}{\sqrt{2}}\exp\left(-\frac{i\pi}{4}\right) - i\pi w W(w), \tag{110}$$

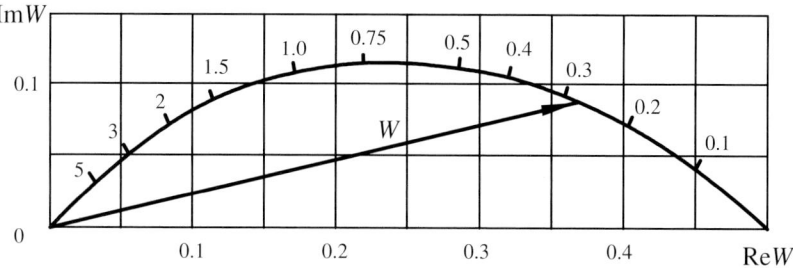

Figure 5. Function $W(w)$ as a vector on a complex plane (ReW, ImW). The graph illustrates values of the argument w (Klem-Musatov, 1980).

and

$$\frac{d^2W(w)}{dw^2} = -i\pi\left[W(w) + w\frac{dW(w)}{dw}\right],\qquad(111)$$

where an overbar denotes a complex conjugate of W.

Figure 5 shows the graph of the function $W(w)$. Note that the derivative of the incomplete gamma function 87 goes to infinity when $z = 0$. However, all derivatives of function $W(w)$ are bounded at $w = 0$.

The function under consideration can be expressed in terms of the Fresnel integral

$$F(z) = \frac{1}{\sqrt{\pi}}\exp\left(-\frac{i\pi}{4}\right)\int_{-\infty}^{z}\exp\left(-ix^2\right)dx.\qquad(112)$$

Substituting the well-known functional relationship (Abramowitz and Stegun, 1972)

$$\frac{1}{2\sqrt{\pi}}\Gamma\left(\frac{1}{2}, z\right) = F\left(-\sqrt{iz}\right) = 1 - F\left(+\sqrt{iz}\right)\qquad(113)$$

into equation 102, we obtain the relationships

$$W(w)\exp\left(\frac{i\pi w^2}{2}\right) = F\left(-\sqrt{\frac{\pi w^2}{2}}\right)\quad\text{and}$$

$$1 - W(w)\exp\left(\frac{i\pi w^2}{2}\right) = F\left(+\sqrt{\frac{\pi w^2}{2}}\right).\qquad(114)$$

Edge diffraction in the boundary layer

Diffraction formulas

Any solution of Kummer's equation 75 can be expressed through a linear combination of its two linearly independent solutions A_1 and A_2,

$$A = C_1' A_1 + C_2' A_2, \tag{115}$$

where C_j' are arbitrary constants. By introducing equations 82 and 105 into equation 115 instead of A_1 and A_2 and taking into consideration equation 52, we obtain

$$
A = \exp(-i\omega\tau)\left\{ C_1' \exp(i\omega\tau_0) \right.
$$
$$
\left. + \operatorname{sign}\alpha \cdot \sqrt{\pi}\, C_2' \left[\frac{1}{2}\exp(i\omega\tau_0) - W(w)\exp(i\omega\tau) \right] \right\}. \tag{116}
$$

Substituting this expression into equation 21, we obtain the solution of the stationary wave equation in the boundary layer

$$f = C_1' f_0 + \delta\sqrt{\pi}\, C_2'\left(\frac{f_0}{2} - f_d\right), \tag{117}$$

$$f_0 = G_0 \exp(i\omega\tau_0), \tag{118}$$

$$f_d = G_0 W(w)\exp(i\omega\tau), \tag{119}$$

$$\delta = \operatorname{sign}\alpha, \tag{120}$$

where the coefficient G_0 satisfies the transport equation 22. By using equations 52 and 101, we can express the quantity w as

$$w = \sqrt{\frac{2iz}{\pi}} = \sqrt{\frac{2i[i\omega(\tau_0 - \tau)]}{\pi}} = \sqrt{\frac{2\omega(\tau - \tau_0)}{\pi}}. \tag{121}$$

To interpret the solution obtained, it is convenient to express the arbitrary constants C_j' through new constants C_j as follows:

$$C_1' = C_1 + \frac{C_2}{2}, \quad C_2' = -\frac{C_2}{\sqrt{\pi}}. \tag{122}$$

Then solution 117 can be written as

$$f = C_1 f_0 + C_2 f',$$
(123)

where

$$f' = s f_0 + \delta f_d, \quad s = \frac{1 - \delta}{2}.$$
(124)

The first item in equation 123 represents the asymptotic ray-theory solution, so consider the second item only. To do this, take $C_1 = 0$, $C_2 = 1$. Then

$$f = f' = s f_0 + \delta f_d = \frac{1 - \delta}{2} f_0 + \delta f_d.$$
(125)

This solution also can be written as

$$f = f_0 - f_d \quad \text{for} \quad \alpha < 0,$$
(126)

$$f = f_d \quad \text{for} \quad \alpha > 0.$$
(127)

There is another useful form of this solution. First consider the case of $\alpha < 0$. By substituting equations 118 and 119 into 126, we obtain

$$
\begin{aligned}
f &= G_0 \big[\exp(i\omega\tau_0) - W(w)\exp(i\omega\tau) \big] \\
&= G_0 \exp(i\omega\tau_0)\big[1 - W(w)\exp[-i\omega(\tau_0 - \tau)] \big] \\
&= f_0 \big[1 - W(w)\exp(-z) \big], \\
z &= i\omega(\tau_0 - \tau).
\end{aligned}
$$
(128)

By expressing the quantity z in terms of w according to equation 101, we obtain

$$f = f_0 \left[1 - W(w)\exp\left(\frac{i\pi w^2}{2} \right) \right] \quad \text{for} \quad \alpha < 0.$$
(129)

In the same way, write equation 127 as

$$f = f_0 W(w)\exp\left(\frac{i\pi w^2}{2} \right) \quad \text{for} \quad \alpha > 0.$$
(130)

Now, using the functional relationships 114, we can represent equations 129 and 130 in the form

$$f = f_0 F\left(+\sqrt{\frac{\pi w^2}{2}}\right) \quad \text{for} \quad \alpha < 0, \tag{131}$$

$$f = f_0 F\left(-\sqrt{\frac{\pi w^2}{2}}\right) \quad \text{for} \quad \alpha > 0. \tag{132}$$

Illuminated and shadow zones

Equations 126 and 127 represent the solution in the form of the sum of two waves: the wave f_0, corresponding to the asymptotic ray-theory equation and the diffracted wave f_d. The amplitude G_0 and the eikonal τ_0 of wave f_0 are determined both for $\alpha < 0$ and for $\alpha > 0$. This follows directly from the description of the diffracted wave 119. However, the wave f_0 itself, as a component of the solution being considered, participates in its description only within domain $\alpha < 0$. Within domain $\alpha > 0$, there is no such wave, and the solution is represented by the diffracted wave only.

This fact can be interpreted in terms of geometric optics. Suppose that wave f_0 is given as a family of rays existing only for $\alpha < 0$. Suppose such rays do not exist for $\alpha > 0$ because of the sharp interruption of the reflection/transmission process caused by the edge. Then there is no wave f_0 within the domain $\alpha > 0$. In such a situation, the region $\alpha < 0$, where the wave f_0 exists, can be called an illuminated zone. The region $\alpha > 0$, where there is no wave f_0, can be called a shadow zone. The common boundary of these regions $\alpha = 0$ can be called a shadow boundary.

The division of space into the illuminated and shadow zones can be taken into account with the following description of the wave:

$$f_0 = \begin{cases} f_0(\tau, \alpha, \beta) & \text{for} \quad \alpha < 0 \quad \text{(illuminated zone)}, \\ 0 & \text{for} \quad \alpha > 0 \quad \text{(shadow zone)}. \end{cases} \tag{133}$$

Quantities G_0 and τ_0 in equation 119 with $\alpha > 0$ can be called the continuation of the amplitude and the eikonal into the shadow zone. The concept of such a continuation will be considered more closely in Chapter 3.

Young's principle

The diffracted wave has different signs on different sides of the shadow boundary; see equations 126 and 127. Let us consider the effect of this fact on the total wavefield when $\alpha \to 0$.

If $\alpha \to 0$, then

$$\tau \to \tau_0, \quad |w| \to 0, \quad W(w) \to W(0) = \frac{1}{2}, \tag{134}$$

and therefore,

$$f_d \to \frac{f_0}{2} \quad \text{as} \quad |\alpha| \to 0. \tag{135}$$

Equations 126 and 127 tend to the following limits:

$$f \to \begin{cases} f_0 - \dfrac{f_0}{2} = \dfrac{f_0}{2} & \text{as} \quad \alpha \to 0 \quad (\alpha < 0), \\[2mm] \dfrac{f_0}{2} & \text{as} \quad \alpha \to 0 \quad (\alpha > 0), \end{cases} \tag{136}$$

or

$$f = \frac{f_0}{2} \quad \text{for} \quad \alpha = 0 \quad (\text{or} \quad \tau = \tau_0). \tag{137}$$

This analysis clearly shows that the continuity of the total wavefield at the shadow boundary $\alpha = 0$ is because of the change of sign of the diffracted wave. This property of the diffracted wave is called the phase inversion. Thus, one can consider the diffraction phenomenon as the effect of superposition of two waves (Figure 6). One of them, f_0, has a discontinuity at its shadow boundary. The other, f_d, can be regarded as a correction that smooths this discontinuity. Such an explanation of the edge-diffraction phenomenon is known in the classical theory of diffraction as Young's principle (Born and Wolf, 2000). Formulas 126 and 127 express this principle in mathematical form.

Boundary layer

The amplitude of the diffracted wave is proportional to the amplitude of the wave f_0. The proportionality factor $W(w)$ characterizes the relative change of the diffracted wave amplitude with the distance from the shadow

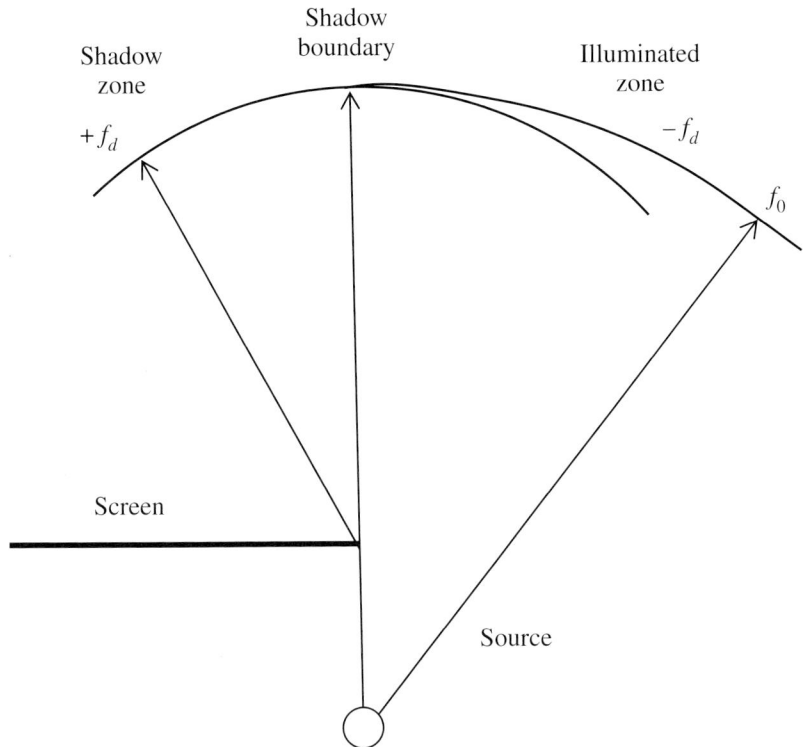

Figure 6. Explanation of the edge-diffraction phenomenon in terms of Young's principle.

boundary. To illustrate the character of this change quantitatively, it is convenient to use a special scale.

Write argument 121 of function $W(w)$ as

$$w = \sqrt{\frac{2\omega(\tau - \tau_0)}{\pi}} = \sqrt{2N}, \tag{138}$$

$$N = \frac{\ell}{\lambda/2}, \quad \ell = c(\tau - \tau_0), \quad \lambda = \frac{2\pi c}{\omega}, \tag{139}$$

where c is a propagation velocity, λ is the wavelength, and N is the phase difference expressed in half wavelengths. The quantity N represents the phase difference in half-period Fresnel zones, so we have expressed quantity w as a function of the number of Fresnel zones. Figure 7 shows the modulus of function $W(w)$ as a function of the number of Fresnel zones. The same function is given in Table 1.

One can see a characteristic feature of the function being considered: Its modulus changes rapidly when $N < 2$ and slowly when $N > 2$. Taking into consideration equation 138, we also can say that $|W|$ changes rapidly when $w < 2$ and slowly when $w > 2$. We also can formulate this fact, choosing as the argument the quantity $(\tau - \tau_0)/T$, where $T = 2\pi/\omega$ is the period of oscillations. To do this, write equation 121 as

$$w = \sqrt{\frac{2\omega(\tau - \tau_0)}{\pi}} = 2\sqrt{\frac{(\tau - \tau_0)}{T}}. \tag{140}$$

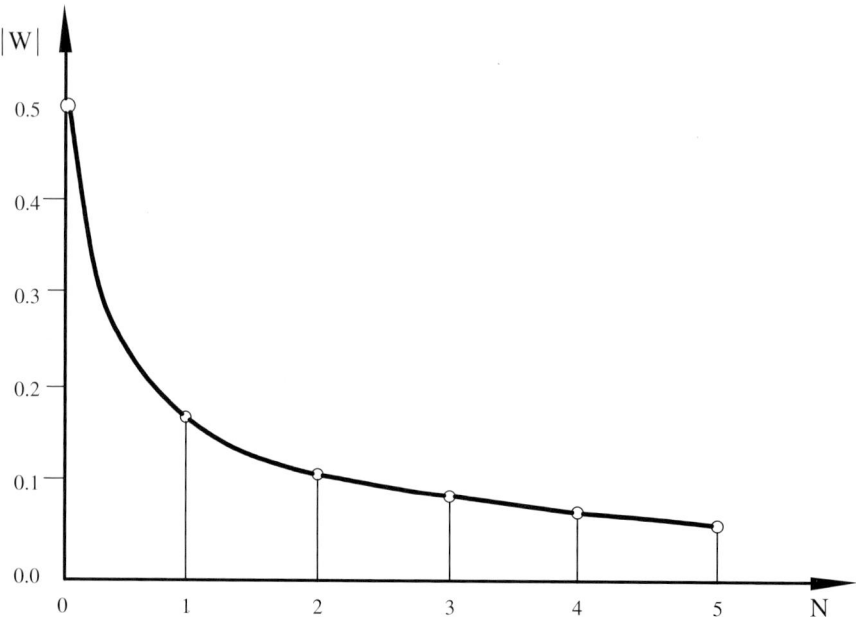

Figure 7. Modulus of W as a function of N, the number of half-period Fresnel zones (Klem-Musatov, 1994).

Table 1. Modulus of function $W(w)$ as a function of the number of Fresnel zones.

N	0	1	2	3	4	5
$|W|$	0.50	0.17	0.11	0.09	0.08	0.07

This expression shows that $w = 2$ when $\tau - \tau_0 = T$. Thus, within the domain of fast change of $|W|$, the phase difference is less than a period of oscillations. Outside this domain, the function $W(w)$ can be described by the asymptotic expression 108.

We can identify a domain of fast change

$$N < 2 \quad \text{or} \quad |w| < 2 \quad \text{or} \quad |\tau - \tau_0| < T \tag{141}$$

with the boundary layer. Comparing inequality 141 in form of $|w^2| < 4$ with inequality 20, we obtain $C = 2\pi$.

Example of a boundary layer

Consider the simplest example of a boundary layer in a 2D homogeneous medium. Let O, S, and M denote the source, diffraction point, and point of observation, respectively. Let r_0, r, and R correspond to distances OS, SM, and OM, respectively (Figure 8). Then

$$\tau = \frac{r_0 + r}{c}, \quad \tau_0 = \frac{R}{c}, \tag{142}$$

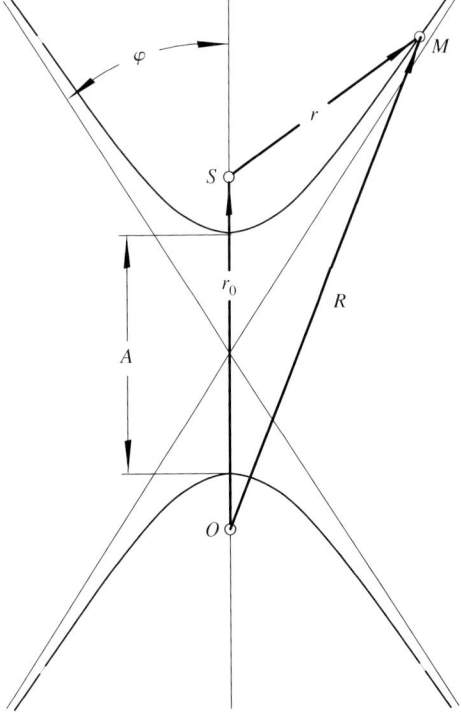

Figure 8. Example of a boundary layer. O is the source, S is the diffraction point, and M is a point of observation. The domain containing point S and bounded by the upper branch of the hyperbola with $w = 2$ forms the boundary layer.

$$w = \sqrt{\frac{2\omega(\tau - \tau_0)}{\pi}} = \sqrt{\frac{2\omega(r_0 + r - R)}{\pi c}} = 2\sqrt{\frac{(r_0 + r - R)}{\lambda}},$$

(143)

where λ is a wavelength.

Consider a line w = constant, at every point of which function $W(w)$ is constant. To do this, we rewrite equation 143 as follows:

$$R - r = A, \quad A = r_0 - \frac{\lambda w^2}{4}.$$

(144)

When w = constant, the quantity A is also constant, so the difference $R - r$ is constant along the line being considered. Recall the definition of a hyperbola as a set of points such that the difference of the distances from each of them to a pair of given points, foci O and S, is a constant A. One can see that equation 144 corresponds to the equation of a hyperbola with foci O and S (Figure 8).

If $R > r$, the line w = constant coincides with that branch of the hyperbola which goes around point S (Figure 9a). The case $R < r$ occurs when point O is a focus (Figure 9b). Then $R < 0$ at point S. In that case, point O corresponds to a sink, not to a source.

The domain containing focus O or S and bounded by hyperbola 144 with $w = 2$, i.e., $A = r_0 - \lambda$, forms the boundary layer. The angle, formed

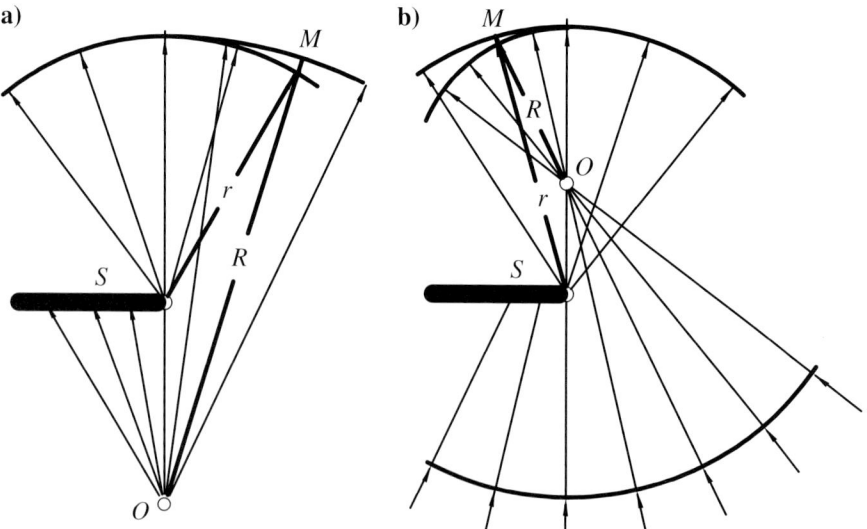

Figure 9. Diffraction of (a) divergent and (b) convergent wavefronts. Point O is the source of the divergent wave and the focus of the convergent wave.

by the asymptotes of the hyperbola with its real axis OS, is determined by the following expression:

$$\varphi = \arctan \sqrt{\left(\frac{r_0}{A}\right)^2 - 1}. \tag{145}$$

The following limits hold true:

$$\lambda = \frac{2\pi c}{\omega} \to 0, \quad A = r_0 - \lambda \to r_0, \quad \varphi \to 0 \quad \text{when} \quad \omega \to \infty. \tag{146}$$

These expressions show that the boundary layer for large values of ω forms a narrow neighborhood near the axis OS, bounded by the hyperbola.

Notice that the above geometric interpretation makes sense only when quantity A is positive. One can see from equation 144 that this is possible if

$$r_0 > \frac{\lambda w^2}{4}. \tag{147}$$

Because we consider the case $|w| < 2$ only, inequality 147 can be rewritten as follows:

$$r_0 > \lambda. \tag{148}$$

It puts a limitation on the minimal distance between points O and S.

Transverse diffusion

To proceed with the physical interpretation of the edge-diffraction phenomenon, we will show how equation 34, with approximation 73, is connected with known equations of asymptotic diffraction theory. To do this, we have to derive a couple of auxiliary relationships.

Express quantity $\Delta(\tau - \tau_0)$ through the divergence of corresponding ray congruences

$$J_0 = \frac{D_0}{c} \quad \text{and} \quad J = \frac{D}{c}, \tag{149}$$

$$D_0 = \frac{D(x, y, z)}{D(\tau_0, \alpha_0, \beta_0)} \quad \text{and} \quad D = \frac{D(x, y, z)}{D(\tau, \alpha, \beta)}, \tag{150}$$

where ray congruences (α_0, β_0) and (α, β) are determined in the section titled "Edge diffraction." Consider the following expression in the boundary layer:

$$\frac{\partial\varphi}{\partial\tau} = \frac{\partial\varphi}{\partial\tau_0}\frac{\partial\tau_0}{\partial\tau},$$

(151)

where φ is an arbitrary function. Represent τ_0 by equation 49:

$$\tau_0(\tau) = \tau + a_2(\tau, 0, \beta)\alpha^2.$$

(152)

It follows from inequality 51 that in the boundary layer, the following asymptotic estimate holds true:

$$\alpha^2 \sim O\left(\frac{1}{\omega}\right) \quad \text{as} \quad \omega \to \infty.$$

(153)

Substituting equation 152 into 151 and taking into consideration estimate 153, we obtain

$$\frac{\partial\varphi}{\partial\tau} = \frac{\partial\varphi}{\partial\tau_0} + O\left(\frac{1}{\omega}\right) \quad \text{as} \quad \omega \to \infty.$$

(154)

This expression shows that it is possible within the boundary layer to replace differentiation with respect to τ_0 by τ. Then from equations 45 and 54, both from Chapter 1, we obtain the desired auxiliary relationships

$$\Delta\tau_0 = \frac{1}{cJ_0}\frac{\partial}{\partial\tau_0}\left(\frac{J_0}{c}\right) = \frac{1}{cJ_0}\frac{\partial}{\partial\tau}\left(\frac{J_0}{c}\right),$$

(155)

$$\Delta\tau = \frac{1}{cJ}\frac{\partial}{\partial\tau}\left(\frac{J}{c}\right),$$

(156)

$$\Delta(\tau - \tau_0) = \frac{1}{cJ}\frac{\partial}{\partial\tau}\left(\frac{J}{c}\right) - \frac{1}{cJ_0}\frac{\partial}{\partial\tau}\left(\frac{J_0}{c}\right).$$

(157)

Represent the solution of equation 34 of this chapter under approximation 73 as

$$A = \frac{\Phi}{G_0},$$

(158)

where G_0 satisfies transport equation 22, i.e.,

$$G_0 = \sqrt{\frac{c}{J_0}}. \tag{159}$$

In the boundary-layer approximation, take

$$G_0 = G_0(\tau, 0, \beta). \tag{160}$$

Substituting expressions 157 through 160 into equation 34 using approximation 73 and taking into consideration equations 44 and 156, we obtain equation

$$\frac{\partial^2 \Phi}{\partial \alpha^2} + \frac{2i\omega g}{c^2} \frac{\partial \Phi}{\partial \tau} + 2i\omega g\,(\Delta\tau)\Phi = 0, \tag{161}$$

$$g = g_{\alpha\alpha}(\tau, 0, \beta). \tag{162}$$

This is a particular case of the so-called parabolic equation, well known in the asymptotic theory of diffraction (Fock, 1965). In contrast with the general case of this equation, the coefficients of equation 161 are fixed at $\alpha = 0$, and there are no derivatives with respect to coordinate β. According to equation 158, equation 161 has the following solution:

$$\Phi = G_0 A, \tag{163}$$

where A is determined by equation 116.

Now take

$$\Phi = GF, \tag{164}$$

where, as in expression 55 of Chapter 1,

$$G = \sqrt{\frac{c}{J}}. \tag{165}$$

Substituting equation 164 into 161 and taking into consideration the boundary-layer approximation

$$G = G(\tau, 0, \beta), \tag{166}$$

we obtain the following equation:

$$\frac{\partial^2 F}{\partial \alpha^2} + \frac{2i\omega g}{c^2}\frac{\partial F}{\partial \tau} + i\omega g\left(\frac{2}{c^2 G}\frac{\partial G}{\partial \tau} + \Delta\tau\right)F = 0. \qquad (167)$$

From expression 56 of Chapter 1, it follows that

$$\frac{2}{c^2 G}\frac{\partial G}{\partial \tau} + \Delta\tau = 0. \qquad (168)$$

Substituting equation 168 into 167, we obtain

$$\frac{\partial^2 F}{\partial \alpha^2} + \frac{2i\omega g}{c^2}\frac{\partial F}{\partial \tau} = 0. \qquad (169)$$

According to equation 164, this equation has the following solution:

$$F = \frac{\Phi}{G}, \qquad (170)$$

where Φ is determined through known functions given by equation 163. From equations 163 and 164, it follows that $A = \Phi/G_0 = GF/G_0$. Introducing this expression into equation 21, we obtain the following representation of the diffracted wave in equations 126 and 127:

$$f_d = GF\exp(i\omega\tau). \qquad (171)$$

This equation describes wave propagation along a congruence of diffracted rays (α, β). Quantity G takes into consideration the geometric spreading of ray tubes in conformity with ray theory. Quantity F takes into consideration the deviation from ray theory. This quantity corresponds to the attenuation function. This function satisfies equation 169, which is an equation of diffusion with an imaginary diffusion coefficient. It is called the equation of transverse diffusion (Fock, 1965) because it describes diffusion of field F in the direction of the coordinate α-line, i.e., along a tangent to the propagating diffracted wavefront (Figure 10).

Thus, one can regard the edge-diffraction phenomenon as the transverse diffusion of energy from the shadow boundary of wave f_0 in the direction tangential to the diffracted wavefront. Because equation 169 does not contain derivatives with respect to β, the diffusion process occurs

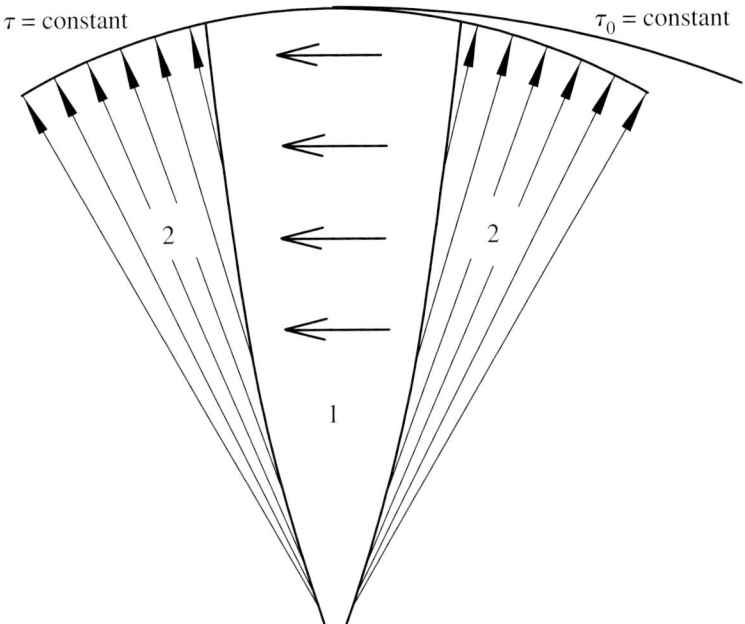

Figure 10. The edge-diffraction phenomenon, shown as transverse diffusion of energy from the shadow boundary of wave f_0 and the associated energy transport from the boundary layer along diffracted ray tubes. Area 1 is the region of transverse diffusion. Areas marked 2 are the regions of energy transport.

independently in each cone of diffracted rays, and there is no redistribution of energy between cones.

    The antisymmetry of the diffracted wave in equations 126 and 127, with respect to coordinate α relative to the shadow boundary, expresses the energy balance in the diffusion process. Inflow of energy into the shadow region (plus sign) is equal to outflow from the illuminated region (minus sign). The diffusion process leads to equalization of amplitudes in the direction tangential to the wavefronts and to smoothing of the shadow boundary. Because the rate of equalization is finite, the processes of energy transport (wave f_0) and transverse diffusion are separated in time. In this case, the process of expansion in time of the region of space encompassed by transverse diffusion has the character of a propagating wave. We can call this the diffracted edge wave and describe it with equation 119. Outside the boundary layer, transverse diffusion dies out and is replaced by an energy-transport mechanism within the scope of the geometric theory of diffraction.

Solution in the neighborhood of a ray

The neighborhood of a ray

Take a congruence of rays (α_0, β_0) with eikonal τ_0 of some wave that can be described within the scope of asymptotic ray theory. Expression 20 determines the neighborhood (or boundary layer) of any surface $\alpha_0 = $ constant formed by the rays of this congruence. Here we will consider a region formed by the intersection of neighborhoods of two surfaces $\alpha_0 = n_1$ and $\alpha_0 = n_2$, where n_j are some numbers. A line of intersection of such surfaces corresponds to some ray $\alpha_0 = $ constant, $\beta_0 = $ constant.

As in the "Edge diffraction" section above, we connect with each of these surfaces its own guide, a curve L, and the additional ray-coordinate system (τ, α, β). We employ the same notations, adding a subscript $j = 1$ or $j = 2$ for the corresponding surface $\alpha_0 = n_j$. Therefore, the guide of the jth surface is denoted as L_j and the ray-coordinate system connected with L_j as $(\tau_j, \alpha_j, \beta_j)$. In these coordinates, surface $\alpha_0 = n_j$ is determined as $\alpha_j = 0$ (Figure 11). The line of intersection of surfaces $\alpha_0 = n_1$ and $\alpha_0 = n_2$ will be determined by a pair of relationships $\alpha_1 = 0$, $\alpha_2 = 0$. The variable of boundary layer 52 will be denoted as

$$z_j = i\omega(\tau_0 - \tau_j), \quad \tau_0 = \tau_0(\tau_j, \alpha_j, \beta_j). \tag{172}$$

Choose surfaces $\alpha_0 = n_1$ and $\alpha_0 = n_2$ with the following condition:

$$\nabla z_1 \cdot \nabla z_2 = 0 \quad \text{for} \quad \alpha_1 = \alpha_2 = 0, \tag{173}$$

where

$$\nabla z_j = (z_j)_{\tau_j} \nabla \tau_j + (z_j)_{\alpha_j} \nabla \alpha_j + (z_j)_{\beta_j} \nabla \beta_j. \tag{174}$$

Bottom indices denote corresponding partial derivatives. From equations 60, it follows that

$$(z_j)_{\tau_j} = (z_j)_{\beta_j} = 0 \quad \text{for} \quad \alpha_j = 0. \tag{175}$$

Therefore, condition 173 is satisfied if

$$\nabla \alpha_1 \cdot \nabla \alpha_2 = 0 \quad \text{for} \quad \alpha_1 = \alpha_2 = 0. \tag{176}$$

This means that surfaces $\alpha_1 = 0$ and $\alpha_2 = 0$ must be mutually orthogonal.

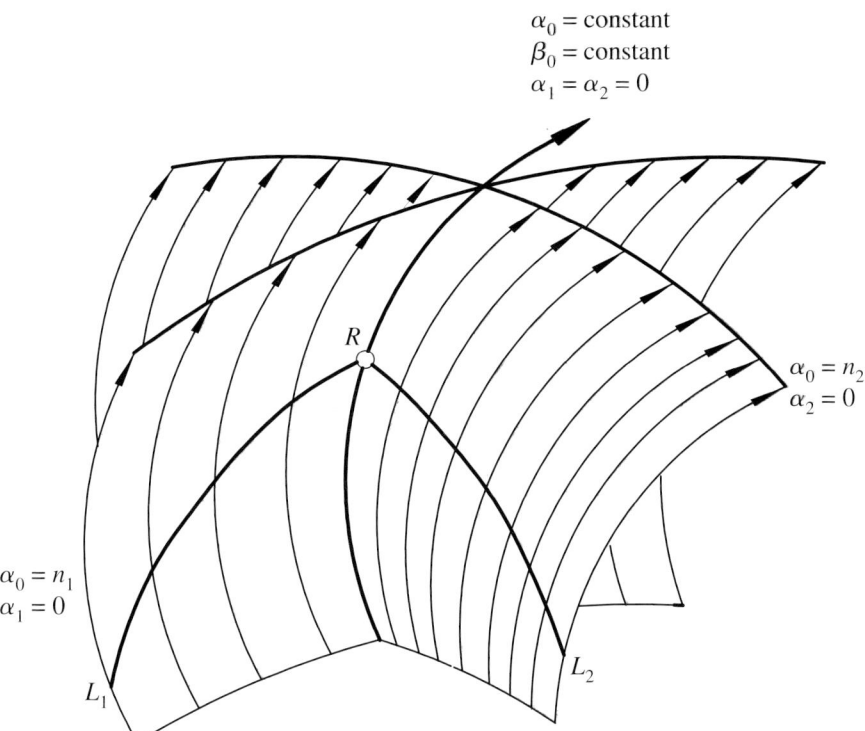

Figure 11. A ray of initial congruence $(\alpha_0,\ \beta_0)$ as the intersection of ray surfaces defined by guide curves L_1 and L_2.

Look for a solution of the stationary wave equation

$$\left(\Delta + k^2\right)f = 0, \quad k = \omega/c \tag{177}$$

in a region formed by the intersection of boundary layers 20

$$\left|\omega\left[\tau_0(M) - \tau_j(M)\right]\right| < C \quad \text{as} \quad \omega \to \infty. \tag{178}$$

Such a region can be called a neighborhood of the ray $\alpha_0 = $ constant, $\beta_0 = $ constant, corresponding to the line of intersection of surfaces $\alpha_1 = 0$ and $\alpha_2 = 0$.

Separation of variables

Look for a solution in the form

$$f = \Phi \exp\left(i\omega\tau_0\right), \tag{179}$$

where Φ is the desired function of a space point. Substituting this expression into equation 177, we obtain an equation for the unknown function Φ as

$$\Delta\Phi + 2i\omega\nabla\tau_0\cdot\nabla\Phi + i\omega\Delta\tau_0\Phi = 0. \tag{180}$$

Introduce the nonorthogonal coordinate system (τ_0, z_1, z_2), where z_j is determined by equation 172, and seek a solution in the form

$$\Phi = T(\tau_0)Z_1(z_1)Z_2(z_2). \tag{181}$$

Here, coefficient T depends only on τ_0 and coefficients Z_j only on z_j. Substituting this expression into equation 180, we obtain the following equation:

$$\begin{aligned}
\Delta TZ_1Z_2 &+ T\Delta Z_1Z_2 + TZ_1\Delta Z_2 \\
&+ 2(\nabla T\cdot\nabla Z_1Z_2 + Z_1\nabla T\cdot\nabla Z_2 + T\nabla Z_1\cdot\nabla Z_2) \\
&+ 2i\omega\nabla\tau_0\cdot(\nabla TZ_1Z_2 + T\nabla Z_1Z_2 + TZ_1\nabla Z_2) \\
&+ i\omega\Delta\tau_0 TZ_1Z_2 = 0.
\end{aligned} \tag{182}$$

On the strength of condition 173, we have

$$\nabla Z_1\cdot\nabla Z_2 = \frac{dZ_1}{dz_1}\frac{dZ_2}{dz_2}\nabla z_1\cdot\nabla z_2 = 0. \tag{183}$$

Taking into consideration this relationship and dividing all items in equation 182 by TZ_1Z_2, we obtain the following equation:

$$\begin{aligned}
\frac{\Delta T}{T} &+ i\omega\left(2\frac{\nabla\tau_0\cdot\nabla T}{T} + \Delta\tau_0\right) \\
&+ \frac{\Delta Z_1}{Z_1} + \frac{2}{Z_1}\left(\frac{\nabla T}{T} + i\omega\nabla\tau_0\right)\cdot\nabla Z_1 \\
&+ \frac{\Delta Z_2}{Z_2} + \frac{2}{Z_2}\left(\frac{\nabla T}{T} + i\omega\nabla\tau_0\right)\cdot\nabla Z_2 = 0.
\end{aligned} \tag{184}$$

The term $T(\tau_0)$ in 181 is a solution at the ray $\alpha_1 = \alpha_2 = 0$. It follows from the theorem of discontinuity propagation that when $\omega \to \infty$, term $T(\tau_0)$ in 181 satisfies the transport equation. Therefore, $T(\tau_0)$ does not depend on ω. Then the following inequalities hold true:

$$\left|2\frac{\nabla\tau_0\cdot\nabla T}{T} + \Delta\tau_0\right| \gg \frac{1}{\omega}\left|\frac{\Delta T}{T}\right| \quad \text{as} \quad \omega \to \infty, \tag{185}$$

$$\left|\nabla\tau_0\right| \gg \frac{1}{\omega}\left|\frac{\nabla T}{T}\right| \quad \text{as} \quad \omega \to \infty. \tag{186}$$

Neglecting the small terms in equation 184, it can be rewritten as

$$L(T) + L(Z_1) + L(Z_2) = 0, \tag{187}$$

where

$$L(T) = 2i\omega\frac{\nabla\tau_0 \cdot \nabla T}{T} + i\omega\Delta\tau_0, \tag{188}$$

$$L(Z_j) = \frac{\Delta Z_j}{Z_j} + 2i\omega\frac{\nabla\tau_0 \cdot \nabla Z_j}{Z_j}, \quad j = 1, 2. \tag{189}$$

This equation admits separation of variables. To do this, rewrite this equation as

$$L(T) = -L(Z_1) - L(Z_2). \tag{190}$$

The left-hand side of this equation does not depend on z_1 and z_2, and the right-hand side does not depend on τ_0. A nonzero solution is possible only if

$$L(T) = \nu, \quad -L(Z_1) - L(Z_2) = \nu, \tag{191}$$

where ν does not depend on τ_0, z_1, and z_2.

Rewrite the second relationship in 191 as

$$-L(Z_1) = L(Z_2) + \nu. \tag{192}$$

The left-hand side of this equation does not depend on z_2, and the right-hand side does not depend on z_1. It is possible only for a nonzero solution if

$$-L(Z_1) = \mu, \quad L(Z_2) + \nu = \mu, \tag{193}$$

where μ does not depend on z_1 and z_2. Thus, equation 187 has the nonzero solution if

$$L(T) = \nu, \quad L(Z_1) = -\mu, \quad L(Z_2) = \mu - \nu, \tag{194}$$

where ν and μ are constants.

Consider the case of $\nu = \mu = 0$. In this case, taking notations 188 and 189 into consideration, we have equations 194 in the form

$$2\nabla\tau_0 \cdot \nabla T + \Delta\tau_0 T = 0, \tag{195}$$

$$\Delta Z_j + 2i\omega\nabla\tau_0 \cdot \nabla Z_j = 0, \quad \text{for} \quad j = 1, 2. \tag{196}$$

Equation 195 is the transport equation 22. Its solution is a wave amplitude within the scope of asymptotic ray theory,

$$T = G_0, \tag{197}$$

where G_0 is determined by equation 159. Therefore, we look for a solution of equation 196.

Reducing to Kummer's equation

Using differential operations

$$\nabla Z_j = \nabla z_j \frac{dZ_j}{dz_j}, \quad \Delta Z_j = \nabla \cdot (\nabla Z_j) = (\nabla z_j)^2 \frac{d^2 Z_j}{dz_j^2} + \Delta z_j \frac{dZ_j}{dz_j}, \tag{198}$$

rewrite equation 196 as

$$(\nabla z_j)^2 \frac{d^2 Z_j}{dz_j^2} + (\Delta z_j + 2i\omega\nabla\tau_0 \cdot \nabla z_j) \frac{dZ_j}{dz_j} = 0. \tag{199}$$

Rewrite the coefficients of this equation in the explicit form. By repeating the technique used in transformations 65 through 68, we obtain

$$2i\omega\nabla\tau_0 \cdot \nabla z_j = (\nabla z_j)^2. \tag{200}$$

Substituting this expression into equation 199, we obtain

$$(\nabla z_j)^2 \frac{d^2 Z_j}{dz_j^2} + \left[\Delta z_j + (\nabla z_j)^2\right] \frac{dZ_j}{dz_j} = 0. \tag{201}$$

From equations 61 through 63 with $\alpha_j \to 0$, it follows that

$$(\nabla z_j)^2 = \frac{4z_j^2 (\nabla\alpha_j)^2}{\alpha_j^2}, \quad \Delta z_j = \frac{2z_j (\nabla\alpha_j)^2}{\alpha_j^2}, \tag{202}$$

where the items turning to zero as $\alpha_j \to 0$ are neglected.

Substituting equations 202 into 201, we obtain

$$z_j \frac{d^2 Z_j}{dz_j^2} + \left(\frac{1}{2} + z_j\right)\frac{dZ_j}{dz_j} = 0. \tag{203}$$

Using the substitution

$$Z_j = Y_j \exp\left(-z_j\right), \tag{204}$$

we can reduce equations 203 to Kummer's equations

$$z_j \frac{d^2 Y_j}{dz_j^2} + \left(\frac{1}{2} - z_j\right)\frac{dY_j}{dz_j} - \frac{1}{2} Y_j = 0. \tag{205}$$

A combination of linearly independent solutions of these equations follows from equations 80 and 100:

$$Y_j = C'_{1j} \exp z_j + C'_{2j}\delta_j \sqrt{\pi}\left[\frac{\exp z_j}{2} - \frac{1}{2\sqrt{\pi}}\Gamma\left(\frac{1}{2}, z_j\right)\exp z_j\right] \tag{206}$$

and

$$\delta_j = \text{sign } \alpha_j, \tag{207}$$

where C'_{1j} and C'_{2j} are arbitrary constants.

Substituting equation 206 into 204, we obtain

$$Z_j = C'_{1j} + C'_{2j}\delta_j \sqrt{\pi}\left[\frac{1}{2} - \frac{1}{2\sqrt{\pi}}\Gamma\left(\frac{1}{2}, z_j\right)\right]. \tag{208}$$

Transform equation 208 to the more convenient form by using equations 102 and 104 and introducing new arbitrary constants as in equations 122:

$$C'_{1j} = C_{1j} + \frac{C_{2j}}{2}, \quad C'_{2j} = -\frac{C_{2j}}{\sqrt{\pi}}. \tag{209}$$

Then equation 208 can be written as

$$Z_j = C_{1j} + C_{2j}\left\{[s_j + \delta_j W(w_j)\exp[i\omega(\tau_j - \tau_0)]\right\}, \tag{210}$$

$$s_j = \frac{1 - \delta_j}{2}, \quad w_j = \sqrt{\frac{2\omega(\tau_j - \tau_0)}{\pi}}. \tag{211}$$

Diffraction formulas

Substituting expressions 197 and 210 into equations 181 and 179, we obtain the desired solution

$$f = C_1 f_0 + C_2 (s_1 f_0 + \delta_1 f_{d1}) + C_3 (s_2 f_0 + \delta_2 f_{d2}) + C_4 f', \tag{212}$$

$$f_0 = G_0 \exp(i\omega\tau_0), \tag{213}$$

$$f_{dj} = G_0 W(w_j) \exp(i\omega\tau_j), \tag{214}$$

$$f' = s_1 s_2 f_0 + s_2 \delta_1 f_{d1} + s_1 \delta_2 f_{d2} + \delta_1 \delta_2 f_R, \tag{215}$$

$$f_R = G_0 W(w_1) W(w_2) \exp(i\omega\tau_R), \tag{216}$$

$$\tau_R = \tau_1 + \tau_2 - \tau_0, \tag{217}$$

$$w_j = \sqrt{\frac{2\omega(\tau_j - \tau_0)}{\pi}}, \tag{218}$$

$$\delta_j = \operatorname{sign} \alpha_j, \tag{219}$$

and

$$s_j = \frac{1 - \delta_j}{2}, \tag{220}$$

where new notations for the arbitrary constants are introduced:

$$C_1 = C_{11}C_{12}, \quad C_2 = C_{21}C_{12}, \quad C_3 = C_{11}C_{22}, \quad C_4 = C_{21}C_{22}. \tag{221}$$

The first item in equation 212 corresponds to a wave within the scope of asymptotic ray theory. The second and third items describe edge diffraction in corresponding boundary layers. Properties of all these items have been considered in previous subsections. Therefore, take

$$C_1 = C_2 = C_3 = 0, \quad C_4 = 1 \tag{222}$$

and consider the solution corresponding to equation 215:

$$f = s_1 s_2 f_0 + s_2 \delta_1 f_{d1} + s_1 \delta_2 f_{d2} + \delta_1 \delta_2 f_R. \tag{223}$$

To interpret this expression, rewrite it as

$$f = s_2 F_{01} + \delta_2 F_{d2}, \tag{224}$$

$$F_{01} = \begin{cases} f_0 - f_{d1} & \text{for} \quad \alpha_1 < 0, \\ f_{d1} & \text{for} \quad \alpha_1 > 0 \end{cases} \tag{225}$$

and

$$F_{d2} = \begin{cases} f_{d2} - f_R & \text{for} \quad \alpha_1 < 0, \\ f_R & \text{for} \quad \alpha_1 > 0 \end{cases} \tag{226}$$

or in the form

$$f = s_1 F_{02} + \delta_1 F_{d1}, \tag{227}$$

$$F_{02} = \begin{cases} f_0 - f_{d2} & \text{for} \quad \alpha_2 < 0, \\ f_{d2} & \text{for} \quad \alpha_2 > 0 \end{cases} \tag{228}$$

and

$$F_{d1} = \begin{cases} f_{d1} - f_R & \text{for} \quad \alpha_2 < 0, \\ f_R & \text{for} \quad \alpha_2 > 0. \end{cases} \tag{229}$$

Let us begin with the analysis of wave f_0. Its amplitude and eikonal are given by equation 213 for all values of $\alpha_j (j = 1, 2)$. However, wave f_0 itself, as a component of the solution being considered, participates in its description only within the intersection of domains $\alpha_1 < 0$, $\alpha_2 < 0$. There is no such wave outside the intersection. This fact can be interpreted in terms of geometric optics in the same way as discussed in the section titled "Illuminated and shadow zones" in this chapter.

Suppose wave f_0 is given at a family of rays existing only for $\alpha_1 < 0$, $\alpha_2 < 0$. Suppose such rays do not exist outside this domain because of the sharp interruption of the reflection/transmission process caused by the edge. For example, the edge has a break point $\alpha_1 = \alpha_2 = 0$ so that the reflecting/ transmitting interface has the form of a quarter plane. In such a situation, the region where wave f_0 exists can be called the illuminated zone.

The region where no f_0 wave exists can be called the shadow zone. The common boundary of these regions is formed by a pair of semi-infinite surfaces, $\alpha_1 = 0$ with $\alpha_2 < 0$ and $\alpha_2 = 0$ with $\alpha_1 < 0$, which have the line $\alpha_1 = \alpha_2 = 0$ in common. We can say that this common boundary consists of two shadow boundaries, $\alpha_1 = 0$ with $\alpha_2 < 0$ and $\alpha_2 = 0$ with $\alpha_1 < 0$ (see corresponding regions and boundaries in Figure 12).

Separation of space into illuminated and shadow zones can be defined by the following description of the wave being considered:

$$
f_0 = \begin{cases}
f_0(\tau_j, \alpha_j, \beta_j) & \text{for} \quad \alpha_1 < 0, \quad \alpha_2 < 0 \text{ (illuminated zone)}, \\
0 & \text{for} \begin{cases}
\alpha_1 > 0, \quad \alpha_2 < 0 \\
\text{or} \\
\alpha_1 > 0, \quad \alpha_2 > 0 \text{ (shadow zone)} \\
\text{or} \\
\alpha_1 < 0, \quad \alpha_2 > 0,
\end{cases}
\end{cases}
\tag{230}
$$

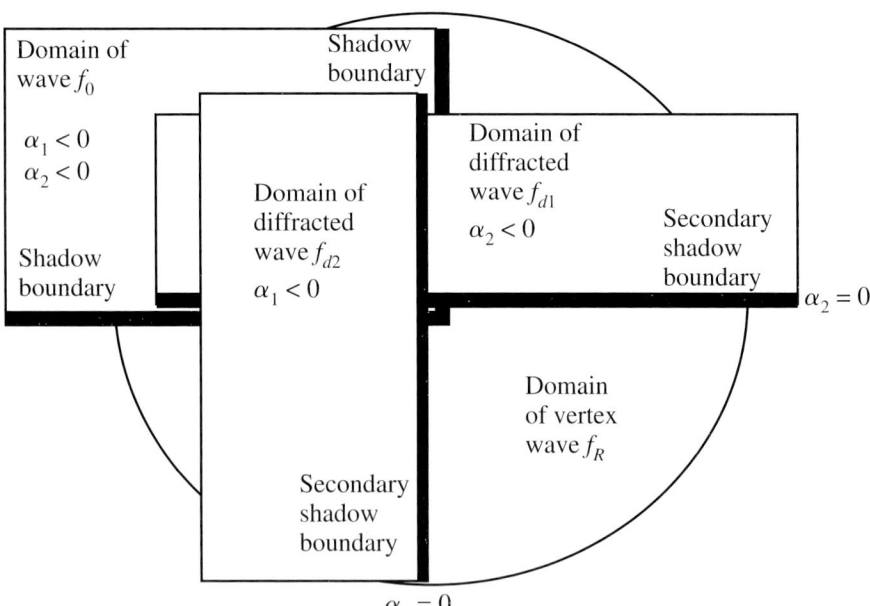

Figure 12. Domains of individual waves with shadow boundaries in the neighborhood of the ray $\alpha_1 = 0$, $\alpha_2 = 0$.

with $j = 1$ or $j = 2$. Quantities G_0 and τ_0 outside the illuminated zone can be called the continuation of the amplitude and the eikonal into the shadow zone.

Secondary illuminated and shadow zones

Equations 225 and 228 illustrate that diffracted waves f_{dj} with $j = 1, 2$ smooth the discontinuities of wave f_0 at its shadow boundaries, $\alpha_1 = 0$ with $\alpha_2 < 0$ and $\alpha_2 = 0$ with $\alpha_1 < 0$. However, waves f_{dj}, as components of the solution being considered, participate in the description only within semi-infinite domains $\alpha_\nu < 0$ with $\nu \neq j$ (wave f_{d1} exists for $\alpha_2 < 0$ and wave f_{d2} for $\alpha_1 < 0$). Outside these domains, there are no waves f_{dj}. This is a specific feature of the considered solution.

This fact again can be interpreted in terms of geometric optics. Suppose wave f_{dj} is given at a congruence of rays (α_j, β_j), existing for $\alpha_\nu < 0$ with $\nu \neq j$ only. Suppose such rays do not exist outside this domain because of the sharp interruption of edge diffraction caused by the break of the edge. A region where wave f_{dj} exists can be called an illuminated zone of this wave or the secondary illuminated zone of the total wavefield. The region where no wave f_{dj} exists can be called the secondary shadow zone. The common boundary of these regions $\alpha_\nu = 0$ with $\nu \neq j$ can be called the secondary shadow boundary (Figure 12).

Separation of space into the secondary illuminated and shadow zones can be explained by the following form describing the diffracted wave:

$$
f_{dj} =
\begin{cases}
f_{dj}(\tau_j, \alpha_j, \beta_j) & \text{for} \quad \alpha_\nu < 0 \\
& \text{and} \quad \nu \neq j \ \text{(the secondary illuminated zone)}, \\
0 & \text{for} \quad \alpha_\nu > 0 \\
& \text{and} \quad \nu \neq j \ \text{(the secondary shadow zone)}.
\end{cases}
\tag{231}
$$

Quantity τ_j outside the region $\alpha_\nu < 0$ with $\nu \neq j$ can be called the continuation of the eikonal into the secondary shadow zone.

Vertex diffraction

Equations 226 and 229 express continuity of the solution at secondary shadow boundaries. Write both equations as

$$
F_{dj} =
\begin{cases}
f_{dj} - f_R & \text{for} \quad \alpha_\nu < 0 \ (\nu \neq j), \\
f_R & \text{for} \quad \alpha_\nu > 0 \ (\nu \neq j),
\end{cases}
\quad \text{for} \ j = 1, 2.
\tag{232}
$$

We will show that this equation expresses the mechanism of smoothing a discontinuity of the diffracted wave at its secondary shadow boundary according to Young's principle. To do this, we substitute equations 214 and 216 into 232 and rewrite the latter as

$$F_{dj} = G_0 W(w_j) F_j, \tag{233}$$

$$F_j = \begin{cases} \exp(i\omega\tau_j) - W(w_\nu)\exp(i\omega\tau_R) & \text{for } \alpha_\nu < 0, \\ W(w_\nu)\exp(i\omega\tau_R) & \text{for } \alpha_\nu > 0. \end{cases} \tag{234}$$

Let us consider some properties of quantity τ_R in this expression as $\alpha_j \to 0$. From equation 50, it follows that

$$\tau_0 - \tau_j = a_{2j}\alpha_j^2, \quad a_{2j} = a_{2j}(\tau_j, 0, \beta_j) = \frac{1}{2}\left(\frac{\partial^2 \tau_0}{\partial \alpha_j^2}\right)_{\alpha_j=0} \quad \text{for } j = 1, 2. \tag{235}$$

By solving this expression for τ_j with $j = 1, 2$ and substituting the results in 217, we have

$$\tau_R = \tau_0 - (a_{21}\alpha_1^2 + a_{22}\alpha_2^2). \tag{236}$$

From equations 235 and 236, we obtain the desired property of this quantity:

$$(\tau_R)_{\alpha_1=0} = \tau_2, \quad (\tau_R)_{\alpha_2=0} = \tau_1. \tag{237}$$

Return to the analysis of equations 233 and 234. Considering equations 237 and 107, we obtain the following limits:

$$\tau_R \to \tau_j, \quad w_\nu \to 0, \quad W(w_\nu) \to \frac{1}{2} \quad \text{as } \alpha_\nu \to 0. \tag{238}$$

Therefore, quantity 234 tends to the following limits:

$$F_j \to \begin{cases} \exp(i\omega\tau_j) - \dfrac{1}{2}\exp(i\omega\tau_j) = \dfrac{1}{2}\exp(i\omega\tau_j) & \text{for } \alpha_\nu < 0, \\ \dfrac{1}{2}\exp(i\omega\tau_j) & \text{for } \alpha_\nu > 0, \end{cases} \quad \text{as } \alpha_\nu \to 0. \tag{239}$$

This shows the continuity of quantity F_j at $\alpha_\nu = 0$. Then in equation 233, we have

$$F_{dj} = \frac{1}{2} G_0 W(w_j) \exp(i\omega\tau_j) = \frac{f_{dj}}{2} \quad \text{for} \quad \alpha_\nu = 0. \quad (240)$$

We can see that the mechanism of smoothing a discontinuity of the diffracted wave at its secondary shadow boundary is the same as the one considered in the section titled "Young's principle" in this chapter. Now, quantity f_R, which changes the sign at the secondary shadow boundary, acts as the smoothing correction. This quantity has the behavior of a wave spreading from the break point of the edge.

Indeed, equation 236 allows us to consider quantity τ_R as a function of three variables $\tau_R = \tau_R(\tau_0, \alpha_1, \alpha_2)$. Fixing two values $\alpha_1 = \text{constant}$ and $\alpha_2 = \text{constant}$, we obtain a line of intersection of corresponding surfaces formed by diffracted rays. Because each surface $\alpha_j = \text{constant}$ ($j = 1, 2$) contains the guideline L_j (the half edge), the line of their intersection passes through the common point R of these curves L_1 and L_2 (the breakpoint of the edge). Therefore, the line of intersection connects the breakpoint of the edge R with a current point of space M (Figure 11). The position of point M on this line is determined by the value of the third variable, $\tau_0 = \text{constant}$. Quantity τ_R satisfies the eikonal equation in the boundary-layer approximation and describes the wave-propagation time along the line. Therefore, we can call quantity f_R the vertex-diffracted wave.

The vertex wave smoothes the discontinuities of both diffracted waves at their secondary shadow boundaries simultaneously (Figure 12). Its amplitude depends on coefficient $W(w_1)W(w_2)$, taking into consideration the diffraction mechanisms that have been considered in the above-mentioned section on Young's principle. However, the mathematical description of vertex diffraction here is so simple only because of the mutual orthogonality of the shadow boundaries.

Criterion for applicability of asymptotic ray theory

Edge and vertex diffraction phenomena limit the range of applicability of asymptotic ray theory. The latter is sufficient in situations in which we can neglect the diffraction components of the total wavefield. To formulate the corresponding criterion, we will analyze the solution of boundary layer 125. It will be clear that the results of such an analysis hold true for the solution in the neighborhood of ray 223.

Take solution 125 in form 131 in the illuminated zone

$$\frac{f}{f_0} = F\left(\sqrt{\frac{\pi w^2}{2}}\right), \quad w = \sqrt{\frac{2\omega(\tau - \tau_0)}{\pi}}. \tag{241}$$

If

$$|w| > 2, \tag{242}$$

in accordance with equation 108, we have

$$\frac{f}{f_0} = 1 + O\left(\frac{1}{\sqrt{\omega}}\right) \quad \text{as} \quad \omega \to \infty, \tag{243}$$

and we can neglect the diffraction component. Thus, asymptotic ray theory is sufficient for description of the wave-propagation process

$$f = f_0. \tag{244}$$

Condition 242 admits a simple geometric interpretation. Quantity w is a function of three points

$$w = w(O, S, M), \tag{245}$$

where O is the source, S is the diffraction point, and M is the observation point. Choose a ray by fixing the positions of points O and M. Then quantity 245 is a function of point S. Expression

$$|w(O, S, M)| = 2 \tag{246}$$

describes a surface in 3D space. Inequality

$$|w(O, S, M)| < 2 \tag{247}$$

determines a region bounded by surface 246 and containing ray OM. Such a region can be called a neighborhood of ray OM. It is easy to see that this definition is equivalent to the corresponding definition in the section titled "Edge diffraction" at the beginning of this chapter.

The criterion for applicability of asymptotic ray theory can be formulated as follows: The solution of the wave equation at ray OM can be described within the scope of asymptotic ray theory 73 if diffraction point S does not belong to the neighborhood of the ray.

A part of the interface belonging to the neighborhood of the ray can be called the neighborhood of the reflection/transmission point. From the above criterion, it follows that the reflection/transmission process can be described within the scope of asymptotic ray theory, if the interface is regular in the neighborhood of the reflection/transmission point.

Consider the simplest example of the neighborhood of a ray in a homogeneous medium by using the formulas in the section titled "Example of a boundary layer" in this chapter. Let formulas 142 and 143 hold true. However, now we write equation 144 as

$$r_0 + r = A', \quad A' = R + \frac{\lambda w^2}{4}. \tag{248}$$

Because points O and M are fixed, quantity R is constant. Because $w = $ constant, quantity A' is also constant. Therefore, sum $r_0 + r$ is constant on the line $w = $ constant. Recall the definition of an ellipse as a set of points so that the sum of distances from each of them to a pair of given points, foci O and M, is a constant A'. We can see that equation 248 corresponds to the equation of an ellipse with foci O and M (Figure 13).

In accordance with the above definition, the neighborhood of a ray corresponds to a region bounded by the ellipse with $w = 2$.

The angle, formed by the diagonal of the ellipse with its real axis OM, is determined by the following expression:

$$\varphi = \arctan\sqrt{1 - \left(\frac{R}{A'}\right)^2}. \tag{249}$$

The following limits hold true:

$$\lambda = \frac{2\pi c}{\omega} \to 0, \quad A' \to R, \quad \varphi \to 0 \quad \text{as} \quad \omega \to \infty. \tag{250}$$

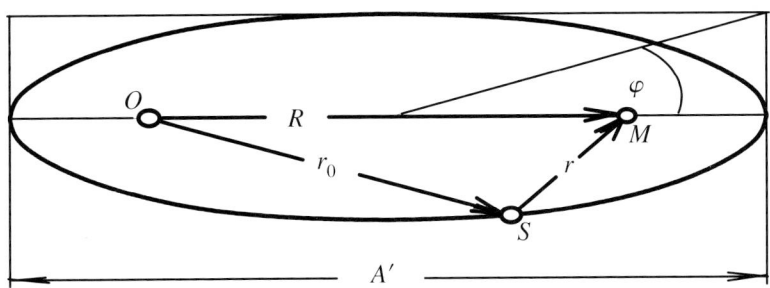

Figure 13. Neighborhood of ray OM as the domain bounded by an ellipse with $w = 2$.

These expressions show that the neighborhood of the ray for large values of ω is a small neighborhood of the ray *OM*, bounded by the ellipse.

Figure 14 shows an example of the neighborhood of a ray for a three-layer seismic model of a region in western Siberia. The predominant frequency of oscillations is assumed to be 10 Hz.

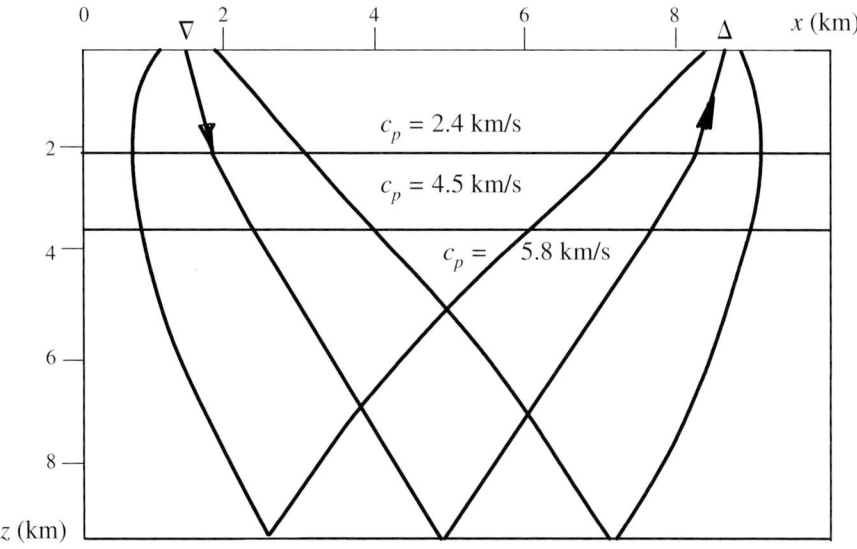

Figure 14. Example of the neighborhood of a ray for a three-layer seismic model (Klem-Musatov, 1980).

Chapter 3

Edge Waves in Boundary Layers

The problem of smoothing a discontinuity at a shadow boundary

The descriptions of edge diffraction that one can use to solve practical problems depend on the character of the problem considered. In some cases, consideration of an individual edge wave can be of practical importance. It is often sufficient in such situations to describe the edge wave in the framework of the geometric theory of diffraction. Sometimes it is necessary to use a more complicated description, which involves combining the formulas of the geometric theory of diffraction and the boundary-layer approximation. However, of greatest practical importance is the case in which edge waves can be regarded as factors interfering with regular reflections/transmissions representing basic geophysical information. In such situations, it is possible to use the simplest description of edge-diffraction phenomena. All the following sections deal only with that kind of situation.

The wavefield pattern considering single diffractions

Let us begin with general considerations on the representation of wavefields in media consisting of regions and interfaces. Description of a stationary wavefield in such media is based on separation of the wavefield into individual waves caused by the consecutive reflection/transmission phenomena at the interfaces. It can be written in the form of the superposition of individual waves:

$$f = \sum_m f_m, \quad f_m = \Phi_m \exp(i\omega\tau_m), \quad (\nabla\tau_m)^2 = \frac{1}{c_m^2}, \tag{1}$$

where m is the index of the individual wave f_m, Φ_m is its amplitude, τ_m is its eikonal, c_m is the propagation velocity, and ω is the frequency of oscillations.

If there are diffracting edges at the interfaces, this description is not sufficient because of shadow zones in the individual wavefields (asymptotic ray theory yields $f_m = 0$ in these zones). We shall denote a singly connected surface dividing the illuminated and shadow zones of an individual wave f_m, i.e., the individual shadow boundary, with the double index mn. Discontinuities of individual waves at the shadow boundaries show the inadequacy of asymptotic ray theory in media with diffracting edges. To smooth such discontinuities, it is sufficient to take the edge waves into consideration. With every mnth shadow boundary of an individual wave f_m, one can connect the corresponding edge wave

$$f_{mn} = \Phi_{mn} \exp\left(i\omega\tau_{mn}\right), \quad \left(\nabla\tau_{mn}\right)^2 = \frac{1}{c_m^2}, \qquad (2)$$

with amplitude Φ_{mn} and eikonal τ_{mn}. Then the equation of the mnth shadow boundary can be given implicitly in the form

$$\tau_{mn} = \tau_m. \qquad (3)$$

By adding the edge waves 2 to each term f_m of sum 1, one can obtain the modified description of the total wavefield in the form

$$f = \sum_m \left(f_m + \sum_n f_{mn} \right). \qquad (4)$$

The total field 4 must be continuous at shadow boundaries. Let us write the condition of continuity in an explicit form. To do this, introduce ray coordinates $(\tau_{mn}, \alpha, \beta)$ of edge wave f_{mn}. Quantities α and β determine a congruence of diffracted rays, i.e., every pair of fixed values $\alpha = $ constant and $\beta = $ constant singles out an individual ray. Let coordinate surface $\alpha = 0$ coincide with the mnth shadow boundary 3 and the illuminated zone of wave f_m coincide with domain $\alpha < 0$. Let quantity β correspond to the angle between the edge and the diffracted ray. Then surface $\beta = $ constant corresponds to a cone of diffracted rays.

One can consider each reflected/transmitted wave as a function of these coordinates:

$$f_m = \Phi_m\left(\tau_{mn}, \alpha, \beta\right)\exp\left[i\omega\tau_m\left(\tau_{mn}, \alpha, \beta\right)\right]. \qquad (5)$$

Then in the neighborhood of the *mn*th shadow boundary, this wave can be written as the discontinuous function

$$f_m = \begin{cases} f_m(\tau_{mn}, \alpha, \beta) & \text{for} \quad \alpha < 0, \\ 0 & \text{for} \quad \alpha > 0. \end{cases} \tag{6}$$

The condition of continuity of sum 4 at the *mn*th shadow boundary of this wave can be written as

$$\left[f_m(\tau_{mn}, \alpha, \beta) + f_{mn}(\tau_{mn}, \alpha, \beta) \right]_{\alpha = -0} = \left[f_{mn}(\tau_{mn}, \alpha, \beta) \right]_{\alpha = +0}. \tag{7}$$

Expression 4, with all items constrained by condition 7, describes the wavefield pattern considering single diffractions. Here, edge waves must act as smoothing corrections that compensate for discontinuities of reflected/transmitted waves at shadow boundaries.

Rational approximation for edge waves

It is possible to describe edge waves in equation 4 by using the simplest approximation that allows us to smooth discontinuities at shadow boundaries and guarantees sufficient accuracy in boundary layers that form shadow-boundary neighborhoods. The following considerations show the rationality of such an approximation: Each original wave f_m in equation 4 is given with some inaccuracy because it corresponds to an asymptotic approximation to the exact solution of the corresponding equation of motion as $\omega \rightarrow \infty$. Therefore, there is no sense in describing the smoothing correction to such a wave with more accuracy than the original wave. The associated description, taking into consideration the inaccuracy of waves f_m and f_{mn}, makes sense when the intensities of these waves are comparable as $\omega \rightarrow \infty$, i.e., in the boundary layers. Outside the boundary layers, the edge waves have the relative order of $O\left(\omega^{-1/2}\right)$ and can be regarded as a diffraction background with respect to the reflected/transmitted wavefields.

These considerations establish the rationale for the boundary-layer approximation for edge waves. Notice that in the framework of such an approximation, diffraction phenomena (which can be described by the geometric theory of diffraction) can be regarded as a diffraction background.

In fact, we already have obtained the corresponding description of edge waves in the boundary-layer approximation, when field 4 was a high-frequency asymptotic solution of the stationary-wave equation. In that case, a wavefield in an individual shadow-boundary neighborhood was given by equations 126 and 127 of Chapter 2, where the diffracted wave 119 of Chapter 2 smoothed the discontinuity of wave 118 of Chapter 2 at its shadow boundary. In this case, we can write the corresponding expression for wave 2 above in form 119 of Chapter 2, changing only the notations. Because the equations of elastodynamics in an isotropic medium as $\omega \to \infty$ can be reduced to scalar wave equations 71 of Chapter 1, diffraction corrections in equation 4 above are known for elastic waves in an isotropic medium as well. In all these cases, there is no need to state the special problem of finding the smoothing corrections because the smoothing effect appears automatically as a property of the corresponding solutions in the boundary layer.

One can use known solutions of equations of motion as smoothing corrections only in sufficiently simple situations. In more general situations, e.g., in the case of nonsmooth edges, one has to state the mathematical problem of obtaining smoothing corrections. Because we will study the case of nonsmooth edges later, we will consider one possible statement of such a problem here.

A specific feature of this statement is that it does not require equations of motion, i.e., their properties must be taken into account only by the kinematics of wavefronts and rays that we suppose to be known. The unique solution of the problem will be singled out by the limitation of a class of possible solutions. This approach is prompted by one of the simplest boundary problems of the theory of analytic functions, which we describe next.

The Sohotsky-Plemelj problem

Here we consider the problem of finding a piecewise-analytic function with a given jump (Dettman, 1965). This problem of singular integral equations is known as the Sohotsky-Plemelj problem (Muskhelishvili, 1953).

Let Γ be a smooth nonself-crossing infinite curve in the complex plane α. Choose the direction on curve Γ as shown by the arrow in Figure 1 and define the left-hand (+) and right-hand (−) sides of this curve. Let $\varphi(\alpha)$ be some function given at Γ. Let this function at Γ be Holder continuous,

$$\left|\varphi(\alpha + \Delta\alpha) - \varphi(\alpha)\right| \leq C|\Delta\alpha|^\mu \quad \text{as} \quad |\Delta\alpha| \to 0, \quad 0 < \mu < 1,$$
(8)

and decreasing at infinity

$$|\varphi(\alpha)| \to 0 \quad \text{as} \quad \alpha \to \infty \ (\alpha \subset \Gamma),$$
(9)

where C is an arbitrary positive constant.

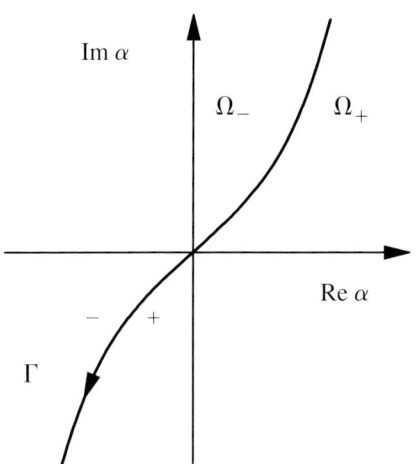

Figure 1. Contour of integration in the Sohotsky-Plemelj problem for complex α.

Let $F(\alpha)$ be an unknown piecewise-analytic function that is discontinuous at Γ. Denote $F^+(\alpha)$ and $F^-(\alpha)$ as the boundary values of this function at left-hand $(+)$ and right-hand $(-)$ sides of contour Γ, respectively. The problem is to find a function $F(\alpha)$ under the following conditions:

$$F^+(\alpha) - F^-(\alpha) = \varphi(\alpha) \quad \text{for} \quad \alpha \subset \Gamma, \tag{10}$$

$$|F(\alpha)| \to 0 \quad \text{as} \quad \alpha \to \infty. \tag{11}$$

This problem has a unique solution given by a Cauchy-type integral,

$$F(\alpha) = \frac{1}{2\pi i} \int_{\Gamma} \frac{\varphi(\eta)}{\eta - \alpha} \, d\eta, \tag{12}$$

where α is a complex variable.

Smoothing the discontinuity at a shadow boundary

We will show that finding the diffraction-correction smoothing for a discontinuity of the reflected/transmitted wave at its shadow boundary can be reduced to the Sohotsky-Plemelj problem.

Represent the amplitude and eikonal of wave 5 in the neighborhood of its shadow boundary $\alpha = 0$ in the boundary-layer approximation by using approximations corresponding to equations 160 and 50 of Chapter 2, given by

$$\Phi_m(\tau_{mn}, \alpha, \beta) \approx \Phi_m(\tau_{mn}, 0, \beta), \tag{13}$$

$$\tau_m(\tau_{mn}, \alpha, \beta) \approx \tau_{mn} + a_2\alpha^2, \tag{14}$$

and

$$a_2 = a_2(\tau_{mn}, 0, \beta) = \frac{1}{2}\left(\frac{\partial^2 \tau_m}{\partial \alpha^2}\right)_{\alpha=0}. \tag{15}$$

Consider the resulting approximate expression as a function of the coordinate α

$$f_m(\alpha) = A \exp(i\omega a_2\alpha^2) \tag{16}$$

and

$$A = A(\tau_{mn}, \beta) = \Phi_m(\tau_{mn}, 0, \beta)\exp(i\omega\tau_{mn}). \tag{17}$$

Variable α in equation 16 is a real quantity. However, one can consider this expression as a description of the function of complex variable $\alpha = x + iy$ along real axis $y = 0$. Indeed, expression 16 with complex values of α gives an analytic function at every finite point in the complex plane of α.

To examine the behavior of this function when $\alpha \to \infty$, separate the real and imaginary parts of the exponent

$$f_m(\alpha) = A \exp\left[-2\omega a_2 xy + i\omega a_2(x^2 - y^2)\right], \quad x = \mathrm{Re}\,\alpha, \quad y = \mathrm{Im}\,\alpha \tag{18}$$

and write the modulus of this function as

$$|f_m(\alpha)| = |A|\exp(-2\omega a_2 xy). \tag{19}$$

One can see that the behavior of quantity 19 in the infinite parts of the complex plane is determined by expressions

$$|f_m(\alpha)| \to 0 \quad \text{as} \quad \alpha \to \infty, \quad \text{if} \quad \mathrm{sign}\,(a_2)xy > 0 \tag{20}$$

and

$$|f_m(\alpha)| \to \infty \quad \text{as} \quad \alpha \to \infty, \quad \text{if} \quad \mathrm{sign}\,(a_2)xy < 0. \tag{21}$$

To find sign (a_2), rewrite equation 14 as

$$a_2 \alpha^2 = \tau_m - \tau_{mn}. \tag{22}$$

Quantity a_2 is a coefficient of the Taylor power-series expansion of the function $\tau_m(\tau_{mn}, \alpha, \beta)$ into the series of form 45 of Chapter 2 over the real variable α. Therefore, in equation 22, the inequality $\alpha^2 \geq 0$ always holds true. Then from equation 22, it follows that

$$\text{sign}(a_2) = \text{sign}(\tau_m - \tau_{mn}). \tag{23}$$

It follows from equations 20 and 23 that the function under consideration is bounded in the infinite parts of the complex α plane determined by expressions

$$xy > 0 \quad \text{for} \quad \tau_m > \tau_{mn}, \quad xy < 0 \quad \text{for} \quad \tau_m < \tau_{mn}. \tag{24}$$

One can see that reflected/transmitted wave 16 can be continued into the complex plane of variable α, including its infinite regions 24. Let us show how the condition of continuity 7 can be expanded into the complex plane. To do this, we consider a curve Γ in the complex plane of α (Figure 1). Then we introduce the analytic function on Γ,

$$f_m(\alpha) = A \exp(i\omega a_2 \alpha^2) \quad \text{for} \quad \alpha \subset \Gamma, \tag{25}$$

where quantities a_2 and A are determined by equations 15 and 17, respectively. This function along contour Γ is Holder continuous (equation 8),

$$|f_m(\alpha + \Delta\alpha) - f_m(\alpha)| \leq C|\Delta\alpha|^\mu \quad \text{for} \quad |\Delta\alpha| \to 0, \quad 0 < \mu < 1, \tag{26}$$

and satisfies condition 9 at infinity

$$|f_m(\alpha)| \to 0 \quad \text{for} \quad |\alpha| \to \infty \quad (\alpha \subset \Gamma). \tag{27}$$

Now look at the unknown diffraction correction in equation 4 as a piecewise-analytic function $f_{mn}(\alpha)$, satisfying conditions

$$f_{mn}^+(\alpha) - f_{mn}^-(\alpha) = f_m(\alpha) \quad \text{for} \quad \alpha \subset \Gamma \tag{28}$$

and

$$|f_{mn}(\alpha)| \to 0 \quad \text{as} \quad |\alpha| \to \infty, \tag{29}$$

where f_{mn}^{+} and f_{mn}^{-} are its boundary values at the right-hand and left-hand sides of contour Γ, respectively. One can see that condition 7 at the shadow boundary is a particular case of condition 28 when $\operatorname{Im}\alpha = 0$.

Thus, if coordinate α is complex, finding the smoothing correction in equation 4 can be reduced to the Sohotsky-Plemelj problem. The latter has the unique solution

$$f_{mn}(\alpha) = \frac{1}{2\pi i} \int_{\Gamma} \frac{f_m(\eta)}{\eta - \alpha} d\eta, \tag{30}$$

where α is a complex quantity.

The case of real variable α

Expression 25 at the real axis coincides with representation of the reflected/transmitted wave in its shadow-boundary neighborhood 6. This means we can describe the reflected/transmitted wave in the neighborhood of its shadow boundary as the piecewise-analytic function 25 at the real axis. Condition 28 at the real axis coincides with the condition of continuity at shadow boundary 7. This means that function 30 of the real variable α smooths the discontinuity of the reflected/transmitted wave at the shadow boundary. We will demonstrate the mechanism of smoothing for the case of $\tau_{mn} > \tau_m$ (the case of $\tau_{mn} < \tau_m$ can be analyzed in the same way).

Let variable α in integral 30 be real and positive. Then in accordance with equation 6, there is no wave f_m, and integral 30 describes quantity f_{mn} in the shadow zone. By means of a continuous change of variable α, we can make it negative. While changing α, we can deform the contour of integration simultaneously in such a way that it does not cross over the pole $\eta = \alpha$. Integral 30 then is transformed into integral

$$F_{mn}(\alpha) = \frac{1}{2\pi i} \int_{\Gamma'} \frac{f_m(\eta)}{\eta - \alpha} d\eta \tag{31}$$

along the new contour Γ' going around the pole $\eta = \alpha$, as illustrated in Figure 2. An integral of an analytic function does not change its value in the deformation of its contour if the singular points of the integrand are not crossed in this process. Therefore, integral 31 describes an analytic function that is continuous with all its derivatives at the shadow boundary.

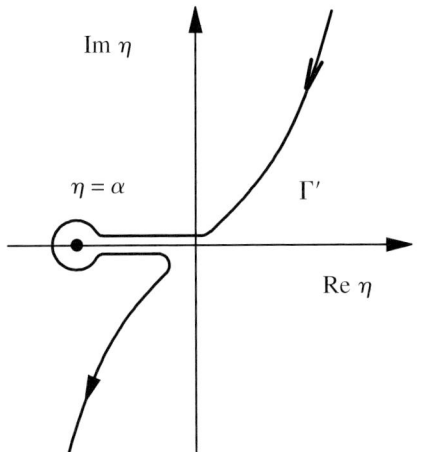

Figure 2. Contour of integration in the solution of the Sohotsky-Plemelj problem for real α.

Notice that the integration along the real axis in Figure 2 has been performed in opposite directions. This means that the contributions to the integral along corresponding parts of the contour of integration cancel each other. Therefore, expression 31 can be rewritten as

$$F_{mn}(\alpha) = f_{mn}(\alpha) + \frac{1}{2\pi i} \int \frac{f_m(\eta)}{\eta - \alpha} d\eta, \tag{32}$$

where the first term corresponds to integral 30 along contour Γ and the second term corresponds to the integral along the closed contour going around the pole at $\eta = \alpha$ counterclockwise. Using the formula for residue at the simple pole

$$\operatorname*{Res}_{\eta=\alpha} \frac{1}{2\pi i} \int \frac{\varphi(\eta)}{\psi(\eta)} d\eta = \frac{\varphi(\alpha)}{\psi'(\alpha)}, \quad \psi' = \frac{d\psi}{d\eta}, \tag{33}$$

we obtain

$$\frac{1}{2\pi i} \int \frac{f_m(\eta)}{\eta - \alpha} d\eta = f_m(\alpha). \tag{34}$$

Because integrals 30 and 31 coincide when $\alpha > 0$, we can rewrite expressions 30 through 34 as

$$F_{mn}(\alpha) = \begin{cases} f_m(\alpha) + f_{mn}(\alpha) & \text{for} \quad \alpha < 0, \\ f_{mn}(\alpha) & \text{for} \quad \alpha > 0, \end{cases} \tag{35}$$

where function $f_m(\alpha)$ is determined by equations 6 and 16 and function $f_{mn}(\alpha)$ is determined by equation 30 for real values of variable α. Expression 35 is equivalent to integral 31, which is an analytic function of the real variable α in the neighborhood of point $\alpha = 0$. Each item in equation 35 has a bounded discontinuity at this point. On the strength of condition 28, function $f_{mn}(\alpha)$ acts as a correction smoothing the discontinuity of function $f_m(\alpha)$.

The smoothing correction as a solution of equations of motion

The Cauchy-type integral as a superposition of solutions of equations of motion

In equations 30 and 31, we introduced the new variable of integration $z = \eta - \alpha$ with real values of α. Then domains where the integrand is an analytic function can be written as

$$(\operatorname{Re} z + \alpha)\operatorname{Im} z > 0 \quad \text{with} \quad \tau_m > \tau_{mn},$$
$$(\operatorname{Re} z + \alpha)\operatorname{Im} z < 0 \quad \text{with} \quad \tau_m < \tau_{mn}. \tag{36}$$

The integrals are transformed to the form

$$f_{mn}(\alpha) = \frac{1}{2\pi i} \int_L \frac{f_m(\alpha + z)}{z}\, dz, \tag{37}$$

$$F_{mn}(\alpha) = \frac{1}{2\pi i} \int_{L'} \frac{f_m(\alpha + z)}{z}\, dz, \tag{38}$$

where contours of integration go to infinity within domains 36 and cross over the real axis in the direction from the upper half plane to the lower half plane. Contour L passes point $z = -\alpha$. Contour L' goes around the left side of the point $z = 0$.

Each of these integrals can be regarded as a superposition of functions $f_m(\alpha)$ depending on the complex parameter z. Let function $f_m(\tau_{mn}, \alpha, \beta)$ satisfy some linear differential equation with coefficients $a_n(\tau_{mn}, \alpha, \beta)$ fixed at $\alpha = 0$. Then the superposition of functions $f_m(\tau_{mn}, \alpha + z, \beta)/z$ over parameter z, on the strength of the linearity of such an operation, satisfies the same equation.

Formal dependence of the positions of contours of integration in equations 37 and 38 on variable α does not matter because contours of integration can be deformed without changing the values of integrals of analytic functions. If field f_m in region $\alpha < 0$ satisfies some linear equation of motion, then integral 37 satisfies the same equation in each individual region $\alpha < 0$ and $\alpha > 0$ but has a discontinuity at $\alpha = 0$. In addition, integral 38 (being an analytic function of α) satisfies the same equation in the union of regions $\alpha < 0$ and $\alpha > 0$.

Thus, in the framework of this approach, there is no need to consider the equations of motion — the smoothing corrections satisfy the corresponding equations of motion automatically.

The smoothing correction as a boundary-layer approximation

We will show here that integral 30 corresponds to a solution of the wave equation in the boundary-layer approximation.

By substituting equation 16 into 30, we have

$$f_{mn} = AI, \tag{39}$$

$$I = \frac{1}{2\pi i} \int_{\Gamma} \frac{\exp\left(i\omega a_2 \eta^2\right)}{\eta - \alpha} d\eta, \tag{40}$$

where quantities a_2 and A are determined by equations 15 and 17, respectively.

Deform the contour of integration as shown in Figure 3 and introduce a new variable of integration t in the following manner:

$$\eta = i\alpha\sqrt{t}. \tag{41}$$

To make this transformation unambiguous, the branch cut in the complex t plane is chosen along the real positive semiaxis, and $\sqrt{t}$ is defined such that

$$\operatorname{Re}\sqrt{t} > 0 \quad \text{for} \quad \arg(t) = 0, \quad \text{if} \quad \alpha < 0,$$
$$\operatorname{Re}\sqrt{t} < 0 \quad \text{for} \quad \arg(t) = 0, \quad \text{if} \quad \alpha > 0. \tag{42}$$

Then the contour of integration Γ is transformed to contour T going around the cut, as illustrated in Figure 4. Integral 40 is transformed to the form

$$I = \frac{1}{4\pi} \int_T \frac{\exp(-zt)}{it - \sqrt{t}} \, dt, \tag{43}$$

and

$$z = i\omega a_2 \alpha^2, \tag{44}$$

where the sign of $\sqrt{t}$ is determined by conditions 42.

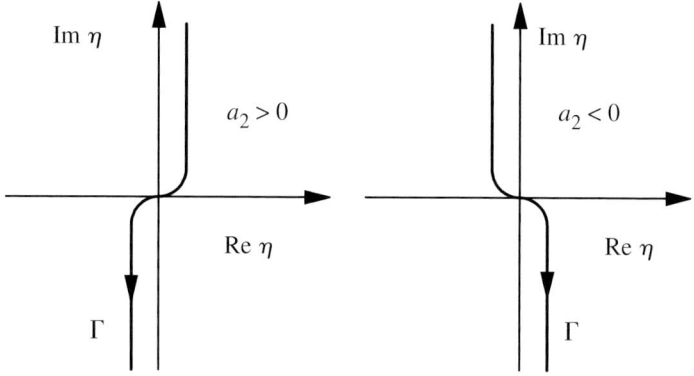

Figure 3. Contours of integration in the convergence domains depending on the sign of a_2.

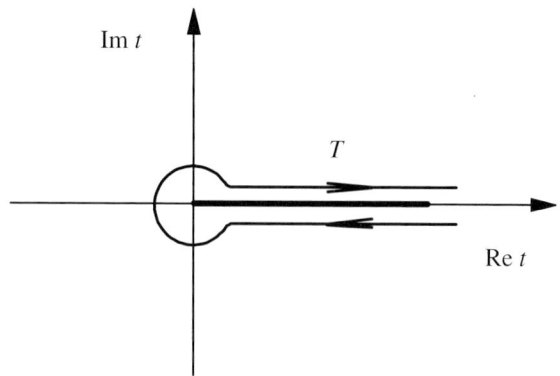

Figure 4. Contour of integration in the complex t plane. The thick line shows the branch cut.

The integral along the arc going around point $t = 0$ tends to zero when the radius of the arc tends to zero. The sign of $\sqrt{t}$ differs on the opposite sides of the cut, in conformity with conditions 42:

$$\sqrt{t} > 0 \quad \text{for} \quad \arg(t) = 0, \quad \sqrt{t} < 0 \quad \text{for} \quad \arg(t) = 2\pi, \quad \text{if} \quad \alpha < 0,$$

$$\sqrt{t} < 0 \quad \text{for} \quad \arg(t) = 0, \quad \sqrt{t} > 0 \quad \text{for} \quad \arg(t) = 2\pi, \quad \text{if} \quad \alpha > 0. \tag{45}$$

As a result, the integral can be regarded as the sum of integrals along the top and bottom sides of the cut. It can be written as

$$I = \frac{1}{4\pi} \int_{\infty}^{0} \frac{\exp(-zt)}{it - \text{sign}(\alpha)\sqrt{t}} dt + \frac{1}{4\pi} \int_{0}^{\infty} \frac{\exp(-zt)}{it + \text{sign}(\alpha)\sqrt{t}} dt$$

$$= \frac{\text{sign}(\alpha)}{2\pi} \int_{0}^{\infty} \frac{\exp(-zt)}{(1+t)\sqrt{t}} dt. \tag{46}$$

The integral obtained can be expressed through special function 102 of Chapter 2. To do this, consider the following combination of the confluent hypergeometric functions:

$$\Psi\left(\frac{1}{2}, \frac{1}{2}; z\right) = \sqrt{\pi}\,\Phi\left(\frac{1}{2}, \frac{1}{2}; z\right) - 2\sqrt{z}\,\Phi\left(1, \frac{3}{2}; z\right). \tag{47}$$

Substituting expressions 80 of Chapter 2 into 47 above, we obtain

$$\Psi\left(\frac{1}{2}, \frac{1}{2}; z\right) = \Gamma\left(\frac{1}{2}, z\right)\exp z, \tag{48}$$

where $\Gamma(1/2, z)$ is the incomplete gamma function. Using relationships 101 and 102 of Chapter 2, we rewrite this expression in the form

$$\Psi\left(\frac{1}{2}, \frac{1}{2}; z\right) = 2\sqrt{\pi}\,W(w), \quad w = \sqrt{\frac{2iz}{\pi}}, \tag{49}$$

where function $W(w)$ is defined by equation 102 of Chapter 2.

On the other hand, function 47 has the known integral representation (Abramowitz and Stegun, 1972)

$$\Psi\left(\frac{1}{2}, \frac{1}{2}; z\right) = \frac{1}{\sqrt{\pi}} \int_0^\infty \frac{\exp(-zt)}{(1 + t)\sqrt{t}} \, dt. \qquad (50)$$

From equations 49 and 50, it follows that

$$\frac{1}{2\pi} \int_0^\infty \frac{\exp(-zt)}{(1 + t)\sqrt{t}} \, dt = W(w), \quad w = \sqrt{\frac{2iz}{\pi}}. \qquad (51)$$

Using equation 51, write integral 46 as

$$I = \text{sign}(\alpha) W(w), \qquad (52)$$

$$w = \sqrt{-\frac{2\omega a_2 \alpha^2}{\pi}}. \qquad (53)$$

Substituting equations 52 and 17 into 39, we obtain

$$f_{mn} = \text{sign}(\alpha) \Phi_m(\tau_{mn}, 0, \beta) W(w) \exp(i\omega\tau_{mn}). \qquad (54)$$

Substituting equation 22 into 53, we obtain

$$w = \sqrt{\frac{2\omega(\tau_{mn} - \tau_m)}{\pi}}. \qquad (55)$$

Comparing expressions 35, 54, and 55 above with expressions 126, 127, 119, and 121 of Chapter 2, one can see that the solution of the problem of smoothing a discontinuity in the class of analytic functions corresponds to the solution of the wave equation in the boundary-layer approximation. The only difference is the description of the reflected/transmitted wave amplitude. In equation 119 of Chapter 2, this amplitude is a function of the current point of space, although in equation 54 above, it is fixed at the shadow boundary. However, this difference does not matter because in the boundary-layer approximation,

$$\Phi_m = \Phi_m(\tau_{mn}, \alpha, \beta) \quad \text{or} \quad \Phi_m = \Phi_m(\tau_{mn}, 0, \beta) \qquad (56)$$

can be chosen arbitrarily. For example, one can write equation 54 as

$$f_{mn} = s_{mn}\Phi_m W(w_{mn}) \exp(i\omega\tau_{mn}), \quad w_{mn} = \sqrt{\frac{2\omega(\tau_{mn} - \tau_m)}{\pi}}, \qquad (57)$$

$$s_{mn} = +1 \quad \text{when} \quad M \subset \Omega_{mn}^+, \quad s_{mn} = -1 \quad \text{when} \quad M \subset \Omega_{mn}^-, \quad (58)$$

where M is an arbitrary point of space and Ω_{mn}^+ and Ω_{mn}^- denote, respectively, the shadow and illuminated zones connected with the mnth shadow boundary of wave f_m. In these expressions, all quantities are functions of the current point of space. The values of $\boldsymbol{\Phi}_m$ and τ_m in the shadow zone must be found by some continuation technique (for example, by analytic continuation). The eikonal can be continued by using expression 14.

Vector waves

Let the terms of sum 4 be vector waves

$$\boldsymbol{f}_m = \boldsymbol{\Phi}_m \exp(i\omega\tau_m), \quad \boldsymbol{\Phi}_m = \mathbf{p}_m \varphi_m, \quad (59)$$

where $\mathbf{p}_m$ is the unit vector of polarization and φ_m is a scalar. In an isotropic medium, vector $\mathbf{p}_m$ coincides with the tangent to the ray (a longitudinal wave) or is orthogonal to it (a transverse wave). Let $\mathbf{j}_1, \mathbf{j}_2, \mathbf{j}_3$ be the unit vectors of a certain fixed coordinate system (for example, a Cartesian system). By decomposing the vector amplitude of wave 59 on this basis, we have

$$\boldsymbol{f}_m = \sum_{q=1}^{3} \mathbf{j}_q f_m^q \quad (60)$$

and

$$f_m^q = \psi_m^q \exp(i\omega\tau_m), \quad (61)$$

where ψ_m^q is a scalar.

We represent the required edge wave $\boldsymbol{f}_{mn}$ in sum 4 as a decomposition on the same basis:

$$\boldsymbol{f}_{mn} = \sum_{q=1}^{3} \mathbf{j}_q f_{mn}^q, \quad f_{mn}^q = \psi_{mn}^q \exp(i\omega\tau_{mn}), \quad (62)$$

where ψ_{mn}^q are unknown scalars. To find each individual scalar function $f_{mn}^q(\tau_{mn}, \alpha, \beta)$, we use the same approach described above for finding the scalar edge wave. This allows us to determine three scalar functions:

$$f_{mn}^q = s_{mn}\psi_m^q W(w_{mn})\exp(i\omega\tau_{mn}) \quad \text{for} \quad q = 1, 2, 3. \quad (63)$$

Inserting equations 63 into 62, we obtain equation 57 again, where quantity Φ_m is now the vector amplitude of wave 59. Druzhinin (1988, 1990) shows that these equations are valid in anisotropic media.

This result must be interpreted. Let $\mathbf{p}_{mn}$ be a unit vector of polarization of vector wave f_{mn}. In accordance with the general theory, this vector must coincide with the tangent to the diffracted ray (a longitudinal wave) or be orthogonal to it (a transverse wave). However, the resulting vector $\mathbf{p}_{mn}$ in vector expression 57 coincides with vector $\mathbf{p}_m(\tau_{mn}, \alpha, \beta)$, which disagrees with the general theory. If quantity Φ_m is fixed at the shadow boundary, the resulting vector $\mathbf{p}_{mn}$ coincides with vector $\mathbf{p}_m(\tau_{mn}, 0, \beta)$. Any of these variants seem to disagree with the general theory. However, all these variants are equal in the boundary-layer approximation

$$\mathbf{p}_{mn}(\tau_{mn}, \alpha, \beta) = \mathbf{p}_{mn}(\tau_{mn}, 0, \beta) = \mathbf{p}_m(\tau_{mn}, 0, \beta) = \mathbf{p}_m(\tau_{mn}, \alpha, \beta),$$
(64)

and the real inaccuracy of the description of polarization does not depend on the choice among them.

The reciprocity principle

Notice one important property of edge wave 57. To find eikonals τ_m and τ_{mn} in the expression for w_{mn}, it is necessary to find two propagation paths from source to observation point. One of them complies with the usual Fermat's principle. The other also complies with this principle, but it includes a point of the edge (the generalized Fermat's principle). Traveltimes along these paths correspond to quantities τ_m and τ_{mn}. These quantities (and hence quantity w_{mn}) do not change if the source is replaced by the observation point and vice versa. Expression 57 does not change its value either if quantity f_m does not change. Therefore, the edge wave complies with the reciprocity principle — it does not change its value when the source is replaced with the observation point and vice versa.

Chapter 4

Tip Waves

The condition of continuity at the secondary shadow boundary

Secondary shadow boundaries

Interfaces in 3D block media are represented by the surfaces of curvilinear polyhedrons. Such surfaces include vertices where several diffracting edges can converge. This means that edges are not smooth in 3D block media. Each smooth part of the edge has two terminal points called tips. A joint of two tips is called a breakpoint of the edge. A joint of two or more tips is a vertex. The existence of such points puts some limitations on the previous discussion of edge diffraction. The essence of these limitations becomes clear from the following example.

Figure 1a shows the geometry of the shadow boundary of a reflected/transmitted wave in the case of a broken edge. The edge consists of two semi-infinite parts, *RA* and *RB*. Point *R* is a breakpoint because the tangent to the edge is not single valued. The shadow boundary consists of two smooth parts, *RAT* and *RBT*. Figure 1b and 1c shows the edge wavefronts arising at semiedges *RB* and *RA*. The diffracted rays, generated by each individual semiedge, form a congruence. However, the unification of two sets of diffracted rays emanating from both semiedges is not a congruence. Each of these congruences exists only on one side of the cone of diffracted rays spreading from point *R*. Such a cone acts as a shadow boundary of the corresponding edge wave generated by the semiedge. The edge wavefield has a discontinuity because there is no edge wave in the region where diffracted rays do not exist. The invalidity of such a description of edge diffraction appears as wavefield discontinuities.

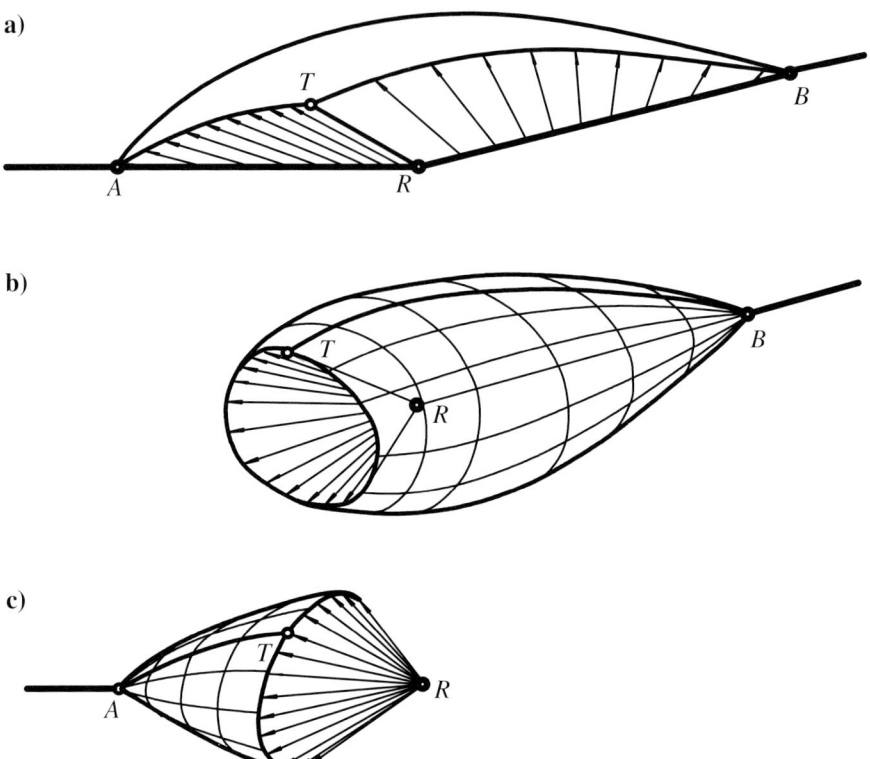

Figure 1. Diffraction on a broken edge *ARB* with a breakpoint *R*. (a) Reflected/transmitted wavefront with the shadow boundaries *BRT* and *ART*. (b) Wavefront diffracted at the edge *RB* with the secondary conic shadow boundary. (c) Wavefront diffracted at the edge *AR* with the secondary conic shadow boundary (Klem-Musatov, 1994).

Diffraction corrections smoothing discontinuities of the edge wavefield can be introduced in the same manner as in Chapter 3. Following the same approach, we first consider such an approach for scalar waves and then generalize the results to vector waves.

First we introduce the necessary definitions. In Chapter 2, we discussed the concepts of illuminated and shadow zones and the shadow boundary of the reflected/transmitted wave, in the section titled "Illuminated and shadow zones." Later in Chapter 2, in the section titled "Secondary illuminated and shadow zones," we modified these concepts for the particular case of a nonsmooth edge. We also introduced concepts of the secondary illuminated and shadow zones and the secondary shadow boundary of edge waves.

From now on, we will use these concepts as bases for the general case. We will speak further of the primary illuminated zone, primary shadow zone, and primary shadow boundary, in relation to the reflected/transmitted wave f_m in expression 4 of Chapter 3. If the edge is not smooth, the region of existence of the edge wave f_{mn} in equation 4 of Chapter 3 is bounded. In this case, an individual edge wave exists within a simply connected region filled with the corresponding congruence of diffracted rays. The wavefield f_{mn} is continuous everywhere within such a region with the exception of the primary shadow boundary $\tau_{mn} = \tau_m$. Such a region is designated as the secondary illuminated zone of wave f_{mn}.

A region where this wave does not exist ($f_{mn} = 0$) is called the secondary shadow zone. The secondary shadow boundary is the simply connected surface dividing these zones. The latter corresponds to a cone of diffracted rays emanating from the tip. In Figure 1b and 1c, the secondary shadow boundaries correspond to the cones of diffracted rays emanating from point R.

In Chapter 3, the superposition of reflected/transmitted waves including edge diffraction was given as equation 4 of that chapter, assuming that diffracting edges were smooth. Now we generalize that description to the case of nonsmooth edges. It is clear that an individual edge wave can have only two secondary shadow boundaries, which we denote with the triple index mnp. We can connect some diffraction correction, f_{mnp}, with each mnpth secondary shadow boundary of wave f_{mn} in such a way that the total wavefield is continuous at this shadow boundary. Then, instead of equation 4 of Chapter 3, we obtain this modified representation of the total wavefield,

$$f = \sum_m \left[f_m + \sum_n \left(f_{mn} + \sum_p f_{mnp} \right) \right], \tag{1}$$

where the summation over indices of all secondary shadow boundaries is included. Discussion of diffraction phenomena caused by the tips can be reduced to finding quantities f_{mnp}, which smooth discontinuities at the secondary shadow boundaries.

Tip wave

Consider quantity f_{mnp} in the form of a wave emanating from a vertex (or a tip). To do this, we must introduce a corresponding congruence of rays. Because it is impossible to determine the tangent plane to an interface at the vertex unambiguously, Snell's law puts no limitations on directions of the rays arising at the vertex. This is the kinematic law of vertex diffraction (Keller, 1962) — the incident ray generates rays leaving the vertex in all directions (Figure 2).

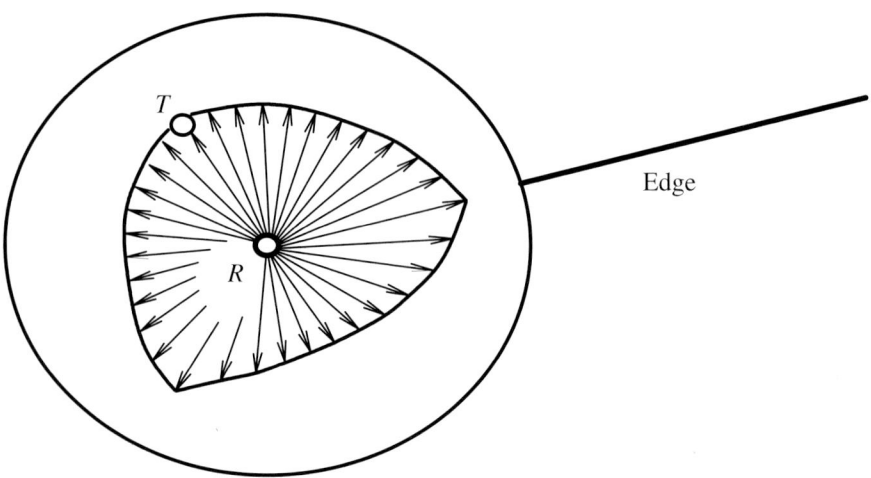

Figure 2. Tip wavefront and rays diverging from a breakpoint R.

Let $\mathbf{e}_{mnp}$ be a unit vector of the tangent to a ray. Let this ray satisfy the kinematic law of vertex diffraction at the tip that results in the mnpth secondary shadow boundary. Then a differential equation of the type of equation 17 of Chapter 1,

$$\frac{d}{ds}\left(\frac{\mathbf{e}_{mnp}}{c_m}\right) = \nabla\left(\frac{1}{c_m}\right), \tag{2}$$

where c_m is the propagation velocity in equation 1 above, determines a congruence of tip-diffracted rays.

Associate a wave with this congruence,

$$f_{mnp} = \Phi_{mnp}\exp\left(i\omega\tau_{mnp}\right), \quad \nabla\tau_{mnp} = \frac{\mathbf{e}_{mnp}}{c_m}, \tag{3}$$

and determine the position of the mnpth secondary shadow boundary by the implicit equation

$$\tau_{mnp} = \tau_{mn}, \tag{4}$$

where τ_{mn} is the eikonal of wave f_{mn} in equation 1. Expression 3 determines a tip-diffracted wave connected with the mnpth secondary shadow boundary.

Domains of continuity

The inclusion of tip waves in equation 1 must guarantee continuity of the total wavefield at secondary shadow boundaries. The tip wave must act as a correction, smoothing a discontinuity at the secondary shadow boundary. The problem of finding such a correction has a difficulty. The edge wave has an additional discontinuity along the secondary shadow boundary because of the phase inversion in equation 58 of Chapter 3. This additional discontinuity does not allow us to formulate the problem of finding the smoothing correction by using the total wavefield's continuity condition at the secondary shadow boundary directly. However, this complication can be avoided by formulating the condition of continuity at the secondary shadow boundary separately in each domain of continuity of the edge wave. To do this, we introduce a special representation of the tip wave, which enables us to formulate the condition in each domain of continuity separately.

Figure 3, illustrating the system of wavefronts $\tau_m = t$, $\tau_{mn} = t$, and $\tau_{mnp} = t$ at a fixed moment t, helps to introduce the concept of domains of continuity of individual waves. Secondary shadow boundary 4 divides

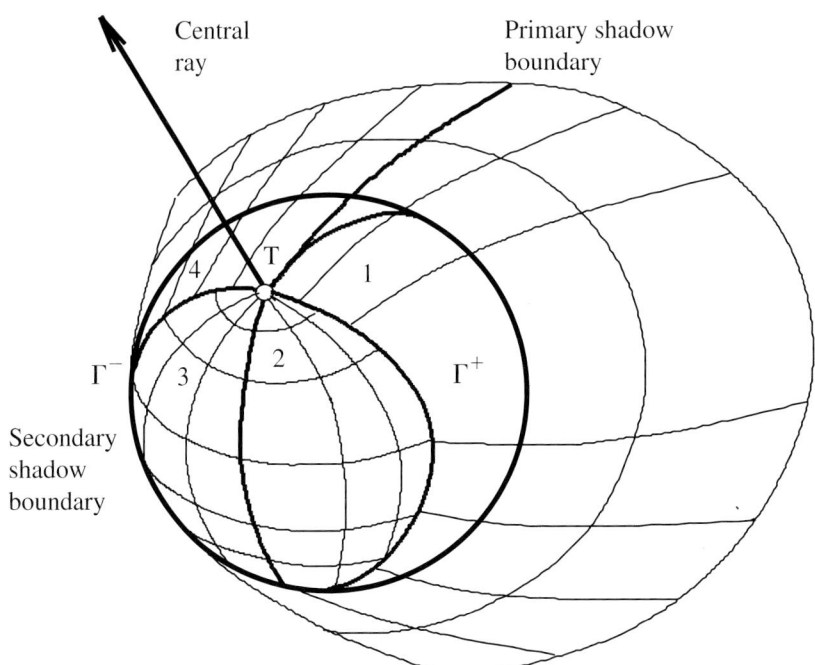

Figure 3. Tip and edge wavefronts connected with an individual secondary shadow boundary in a fixed moment.

space into two parts — the secondary illuminated zone and secondary shadow zone. Suppose that the primary shadow boundary $\tau_{mn} = \tau_m$ can be continued (e.g., analytically) into the secondary shadow zone. Then the continued primary shadow boundary ($\tau_{mn} = \tau_m$) and the secondary shadow boundary (equation 4) divide the domain of tip wave 3 into four parts, depicted in Figure 4. We can call them domains of continuity because within each of them, the edge wave and tip wave are simultaneously continuous. We denote them with numbers 1, 2, 3, and 4, going clockwise around the line $\tau_{mnp} = \tau_{mn} = \tau_m$. The shortest path from the fourth domain to the first coincides with the shortest path from the primary illuminated zone of wave f_m to the primary shadow zone through the mnth primary shadow boundary. We call the line $\tau_{mnp} = \tau_{mn} = \tau_m$ the central ray. Notice that the second and fourth domains have common points only on the same ray.

On crossing the central ray along the secondary shadow boundary, the edge wave f_{mn} has a discontinuity because of the change of sign (equation 58 of Chapter 3). Because of this, the central ray divides the secondary shadow boundary into intervals of continuity of the edge wave. We denote a boundary between the first and second domains by Γ^+ and between the third and fourth domains by Γ^-. The union of surfaces Γ^+ and Γ^- forms the secondary shadow boundary. The edge wavefield is continuous at each individual surface $\Gamma^\pm$.

Suppose that the tip wavefield can be divided into parts

$$f_{mnp} = f^+ + f^-, \quad f^\pm = \Phi^\pm \exp\left(i\omega\tau_{mnp}\right), \tag{5}$$

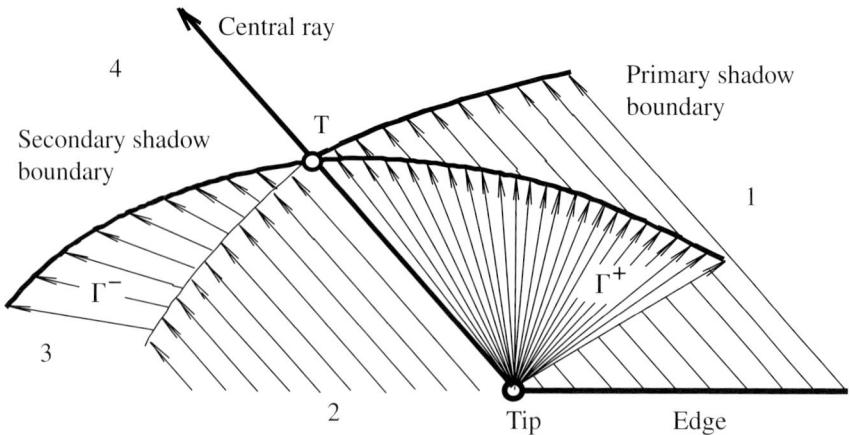

Figure 4. Domains of continuity of tip and edge waves in the neighborhood of the central ray. The union of domains 1 and 4 corresponds to the secondary illuminated zone. The union of domains 2 and 3 corresponds to the secondary shadow zone. Domains 2 and 3 are separated by the analytic continuation of the primary shadow boundary.

so that sum $f_{mn} + f^+$ is continuous on crossing surface Γ^+ and sum $f_{mn} + f^-$ is continuous on crossing surface Γ^-. This representation of the tip wave allows us to formulate the continuity condition appropriately at the secondary shadow boundary for the problem of finding the smoothing correction. To write the continuity condition, it is necessary to introduce the corresponding coordinate system, defined in the next section.

Ray coordinates

Let $\left(\tau_{mnp},\ \psi^\pm,\ \sigma\right)$ be ray coordinates of waves $f^\pm$. Here, $\psi^\pm$ and σ determine a congruence of tip-diffracted rays so that each pair of fixed values, $\psi^\pm = $ constant and $\sigma = $ constant, singles out an individual ray. Coordinate σ corresponds to the distance from the central ray (for example, it can be the arc length on surface $\tau_{mnp} = $ constant or the angle between the tip-diffracted ray and the central ray at the tip). Coordinate $\psi^\pm$ corresponds to the distance from surface $\Gamma^\pm$ and can vary in the interval $-\pi \leq \psi^\pm \leq \pi$. This always can be accomplished with proper scaling. Let surface $\psi^\pm = 0$ coincide with surface $\Gamma^\pm$ and surface $|\psi^\pm| = \pi$ with surface $\Gamma^\mp$ (Figure 5).

Further, we must express values of coordinate $\psi^\pm$ for values less than $\pi/2$. To do this, we introduce special notations for domains of continuity. Let Ω_{mnp}^- denote the union of the first and third domains of continuity and

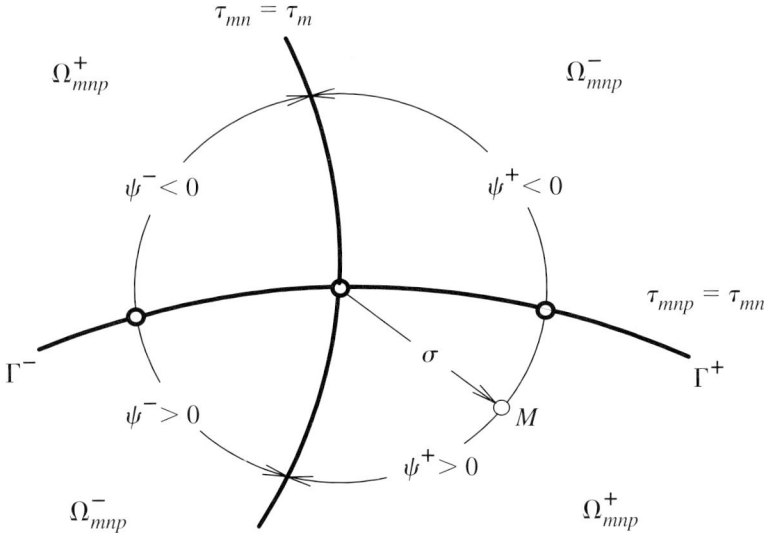

Figure 5. Coordinates $\psi^\pm$ and σ on the tip wavefront $\tau_{mnp} = $ constant (from Figure 14, Klem-Musatov and Aizenberg [1985]). © 1985, *Journal of Geophysics*. Used with kind permission of Springer Science and Business Media.

the union of the second and fourth domains (see Figures 4 and 5). Introduce unit function

$$s_{mnp} = \begin{cases} +1 & \text{for} \quad M \subset \Omega_{mnp}^+, \\ -1 & \text{for} \quad M \subset \Omega_{mnp}^-, \end{cases} \tag{6}$$

where M is a current point in space. This function is connected with unit function 58 of Chapter 3 by the obvious relationships

$$s_{mnp} = \begin{cases} -s_{mn} & \text{for} \quad \psi^\pm < 0 \quad \text{(the secondary illuminated zone)}, \\ +s_{mn} & \text{for} \quad \psi^\pm > 0 \quad \text{(the secondary shadow zone)}. \end{cases} \tag{7}$$

It is easy to check that the following relationships hold true:

$$\psi^\pm = \begin{cases} \pm s_{mnp} |\psi^\pm| & \text{for} \quad |\psi^\pm| \le \pi/2, \\ \pm s_{mnp} (|\psi^\pm| - \pi) & \text{for} \quad |\psi^\pm| > \pi/2. \end{cases} \tag{8}$$

Using these relationships and taking into consideration the relationship

$$|\psi^\pm| + |\psi^\mp| = \pi, \tag{9}$$

we can express $\psi^\pm$ through its value, less than $\pi/2$,

$$\psi^+ = s_{mnp} |\psi^+|, \quad \psi^- = -s_{mnp} (|\psi^+| - \pi)$$
$$\text{for} \quad |\psi^+| < \pi/2 \quad \text{and} \quad |\psi^-| > \pi/2,$$
$$\psi^+ = s_{mnp} (|\psi^-| - \pi), \quad \psi^- = -s_{mnp} |\psi^-|$$
$$\text{for} \quad |\psi^+| > \pi/2 \quad \text{and} \quad |\psi^-| < \pi/2. \tag{10}$$

Edge wave as a function of the tip-wave ray coordinates

We write the edge wave as a function of tip-wave ray coordinates. To do this, we express unit function 58 of Chapter 3 first as a function of coordinate $\psi^\pm$. The intersection of primary shadow zone Ω_{mnp}^+ or primary illuminated zone Ω_{mnp}^- of wave f_m with the secondary illuminated zone can be determined by the following inequalities, respectively,

$$-\pi/2 < \psi^+ < 0 \quad \text{for} \quad M \subset \Omega_{mnp}^+, \quad -\pi/2 < \psi^- < 0 \quad \text{for} \quad M \subset \Omega_{mnp}^-, \tag{11}$$

where M is an arbitrary point of space. From these inequalities, we have the following expression of unit function 58 of Chapter 3:

$$s_{mn} = \pm 1 \quad \text{for} \quad -\pi/2 < \psi^{\pm} < 0. \tag{12}$$

Regard the edge wave in domain Ω^{+}_{mnp} as a function of coordinate ψ^{+} and in domain Ω^{-}_{mnp} as a function of ψ^{-}. Then taking equation 12 into consideration, write equation 57 of Chapter 3 as a function of tip-wave ray coordinates in the secondary illuminated zone:

$$f_{mn}\left(\tau_{mnp}, \psi^{\pm}, \sigma\right) = \pm \Phi_{m} W\left(w_{mn}\right) \exp\left(i\omega\tau_{mn}\right) \quad \text{for} \quad -\pi/2 < \psi^{\pm} \le 0. \tag{13}$$

By continuing this expression through the primary shadow boundary $\psi^{\pm} = -\pi/2$ and taking the phase inversion into consideration, we obtain

$$f_{mn}\left(\tau_{mnp}, \psi^{\pm}, \sigma\right) = \mp \Phi_{m} W\left(w_{mn}\right) \exp\left(i\omega\tau_{mn}\right) \quad \text{for} \quad -\pi \le \psi^{\pm} \le -\pi/2. \tag{14}$$

Expressions 13 and 14 determine the edge wave for $\psi^{\pm} < 0$. There is no edge wave in region $\psi^{\pm} > 0$. Thus we can consider the edge wave as a function of coordinates $\left(\tau_{mnp}, \psi^{+}, \sigma\right)$,

$$f_{mn} = \begin{cases} f_{mn}\left(\tau_{mnp}, \psi^{+}, \sigma\right) & \text{for} \quad -\pi \le \psi^{+} \le 0, \\ 0 & \text{for} \quad 0 < \psi^{+} < \pi, \end{cases} \tag{15}$$

with the discontinuity at surface Γ^{+}, or as a function of coordinates $\left(\tau_{mnp}, \psi^{-}, \sigma\right)$,

$$f_{mn} = \begin{cases} f_{mn}\left(\tau_{mnp}, \psi^{-}, \sigma\right) & \text{for} \quad -\pi \le \psi^{-} \le 0, \\ 0 & \text{for} \quad 0 < \psi^{-} < \pi, \end{cases} \tag{16}$$

with the discontinuity at Γ^{-}.

The continuity condition at a secondary shadow boundary

Consider the terms of sum 5 as functions of corresponding ray coordinates

$$f^{+} = f^{+}\left(\tau_{mnp}, \psi^{+}, \sigma\right) = f^{+}(\psi^{+}), \quad f^{-} = f^{-}\left(\tau_{mnp}, \psi^{-}, \sigma\right) = f^{-}(\psi^{-}), \tag{17}$$

omitting dependence on τ_{mnp} and σ for brevity.

We assume in equation 5 that sum $f_{mn} + f^+$ is continuous when crossing surface Γ^+. To illustrate this condition, write the edge wave in the form 15. Then the corresponding condition can be written as

$$\left[f^+(\psi^+) + f_{mn}(\psi^+) \right]_{\psi^+=0_-} = \left[f^+(\psi^+) \right]_{\psi^+=0_+} \tag{18}$$

where, once again, the dependence of f_{mn} on τ_{mnp} and σ is omitted for brevity. We assume in equation 5 that sum $f_{mn} + f^-$ is continuous when crossing surface Γ^-. Representing the edge wave in the form 16, write the corresponding condition as

$$\left[f^-(\psi^-) + f_{mn}(\psi^-) \right]_{\psi^-=0_-} = \left[f^-(\psi^-) \right]_{\psi^-=0_+}. \tag{19}$$

Combining equations 18 and 19 results in

$$\left[f^\pm(\psi^\pm) + f_{mn}(\psi^\pm) \right]_{\psi^\pm=0_-} = \left[f^\pm(\psi^\pm) \right]_{\psi^\pm=0_+}. \tag{20}$$

Approximations in the neighborhood of the central ray

The neighborhood of the central ray

Here we will use the concept of the neighborhood of the central ray to determine the tip-wave amplitude. We defined the neighborhood of the primary shadow boundary by condition 20 of Chapter 2, which can be written as

$$\left| \omega \left[\tau_m(M) - \tau_{mn}(M) \right] \right| < C \quad \text{as} \quad \omega \to \infty, \tag{21}$$

where M is an arbitrary point in space and arbitrary constant C does not depend on the frequency ω. Define the neighborhood of the secondary shadow boundary by analogy with this condition as

$$\left| \omega \left[\tau_{mn}(M) - \tau_{mnp}(M) \right] \right| < C \quad \text{as} \quad \omega \to \infty. \tag{22}$$

Refer to the intersection of regions 21 and 22 as the neighborhood of the central ray. This definition can be written as a single inequality. For example, let the values of eikonals at point M satisfy the inequalities

$$\tau_{mn} \geq \tau_m, \quad \tau_{mnp} \geq \tau_{mn}. \tag{23}$$

Then conditions 21 and 22 can be rewritten as

$$\omega(\tau_{mn} - \tau_m) < C, \quad \omega(\tau_{mnp} - \tau_{mn}) < C, \quad \text{as} \quad \omega \to \infty. \quad (24)$$

Because each of these inequalities holds true in the neighborhood of the central ray, their sum also holds true:

$$\omega(\tau_{mn} - \tau_m) + \omega(\tau_{mnp} - \tau_{mn}) < 2C, \quad \text{as} \quad \omega \to \infty$$

or, equivalently,

$$\omega(\tau_{mnp} - \tau_m) < 2C, \quad \text{as} \quad \omega \to \infty. \quad (25)$$

Replacing $2C$ with C, because it is arbitrary, rewrite this condition as

$$\omega(\tau_{mnp} - \tau_m) < C, \quad \text{as} \quad \omega \to \infty. \quad (26)$$

It also can be rewritten as

$$\left| \omega \left[\tau_m(M) - \tau_{mnp}(M) \right] \right| < C, \quad \text{as} \quad \omega \to \infty. \quad (27)$$

Consider condition 27 as the definition of the neighborhood of the central ray. The same condition can be obtained easily for situations different from 23.

It follows from condition 27 that $|\tau_m - \tau_{mnp}| \to 0$ when $\omega \to \infty$. Because equality $\tau_m = \tau_{mnp}$ holds true at the central ray only, condition 27 holds true in a narrow region surrounding the central ray. In such a region, it is possible to simplify the description of slowly changing functions by using their local approximations. We will consider such approximations next.

The edge-wave eikonal

Expand function $\tau_{mn}(\tau_{mnp}, \psi^\pm, \sigma)$ in a Taylor series in the variable σ in the vicinity of point $\sigma = 0$, neglecting third- and higher-order powers,

$$\tau_{mn}(\tau_{mnp}, \psi^\pm, \sigma) = \tau_{mn}(\tau_{mnp}, \psi^\pm, 0) + \sigma \left(\frac{\partial \tau_{mn}}{\partial \sigma} \right)_{\sigma=0} + \frac{\sigma^2}{2} \left(\frac{\partial^2 \tau_{mn}}{\partial \sigma^2} \right)_{\sigma=0}. \quad (28)$$

Because $\tau_{mnp} = \tau_{mn} = \tau_m$ when $\sigma = 0$, we have

$$\tau_{mn}\left(\tau_{mnp}, \psi^{\pm}, 0\right) = \tau_{mnp}. \tag{29}$$

Consider the following as the directional derivative,

$$\left(\frac{\partial \tau_{mn}}{\partial \sigma}\right)_{\sigma=0} = \left(\nabla \tau_{mn} \cdot \mathbf{e}_{\sigma}\right)_{\sigma=0}, \tag{30}$$

where $\mathbf{e}_{\sigma}$ is the unit vector of the σ-coordinate line, i.e., the line of intersection of coordinate surfaces $\tau_{mnp} = $ constant and $\psi^{\pm} = $ constant. Because vectors $\nabla \tau_{mn}$ and $\mathbf{e}_{\sigma}$ are mutually orthogonal when $\sigma = 0$, we have

$$\left(\nabla \tau_{mn} \cdot \mathbf{e}_{\sigma}\right)_{\sigma=0} = 0 \tag{31}$$

and

$$\left(\frac{\partial \tau_{mn}}{\partial \sigma}\right)_{\sigma=0} = 0. \tag{32}$$

Substituting equations 29 and 32 into 28, we obtain

$$\tau_{mn}\left(\tau_{mnp}, \psi^{\pm}, \sigma\right) = \tau_{mnp} + \frac{\sigma^2 a_{mn}\left(\psi^{\pm}\right)}{2} \tag{33}$$

and

$$a_{mn}\left(\psi^{\pm}\right) = \left(\frac{\partial^2 \tau_{mn}}{\partial \sigma^2}\right)_{\sigma=0}. \tag{34}$$

This expression approximates the edge-wave eikonal in the neighborhood of the central ray, i.e., for small values of σ. To estimate the order of this value for high frequencies ω, substitute equation 33 into inequality 22. Then

$$\left|\frac{\omega \sigma^2 a_{mn}\left(\psi^{\pm}\right)}{2}\right| < C \quad \text{as} \quad \omega \to \infty, \tag{35}$$

where both the arbitrary constant C and a_{mn} are independent of ω. This asymptotic estimation follows from condition 35,

$$\sigma \sim O\left(\frac{1}{\sqrt{\omega}}\right) \quad \text{as} \quad \omega \to \infty. \tag{36}$$

We want to write the edge-wave eikonal in a form containing an explicit and simple dependence on coordinate $\psi^{\pm}$. Such a representation can be found by using expression 33. Take two points of space

$$M\left(\tau_{mnp}, \psi^{\pm}, \sigma\right) \quad \text{and} \quad M^{*}\left(\tau_{mnp}, \psi^{*}, \sigma\right)$$
$$\text{with} \quad \psi^{*} = \pi/2 \quad \text{or} \quad \psi^{*} = -\pi/2. \tag{37}$$

Point M^{*} can be considered as a projection of the arbitrary point M on the primary shadow boundary (see Figure 6). At point M^{*}, expression 33 becomes

$$\tau_{mn}\left(\tau_{mnp}, \psi^{*}, \sigma\right) = \tau_{mnp} + \frac{\sigma^2 a_{mn}\left(\psi^{*}\right)}{2}. \tag{38}$$

Eliminating $\sigma^2/2$ from equations 33 and 38, we find

$$\tau_{mn} = \tau_{mnp} - \left(\tau_{mnp} - \tau_{mn}^{*}\right)A_{mn}, \tag{39}$$

so that

$$A_{mn} = \frac{a_{mn}\left(\psi^{\pm}\right)}{a_{mn}\left(\psi^{*}\right)}, \tag{40}$$

where

$$\tau_{mn}^{*} = \tau_{mn}\left(\tau_{mnp}, \psi^{*}, \sigma\right). \tag{41}$$

The reflected/transmitted wave eikonal

In deriving equation 39, we did not use any specific properties of the congruence of diffracted rays (for example, we did not use the kinematic law of edge diffraction). Therefore, all considerations can be repeated without any changes for the case of reflected/transmitted waves. Consider the reflected/transmitted wave eikonal as a function of coordinates $\psi^{\pm}$ and σ. Assume that it permits analytical continuation from the primary illuminated zone to any point of the neighborhood of the central ray. Then we get an

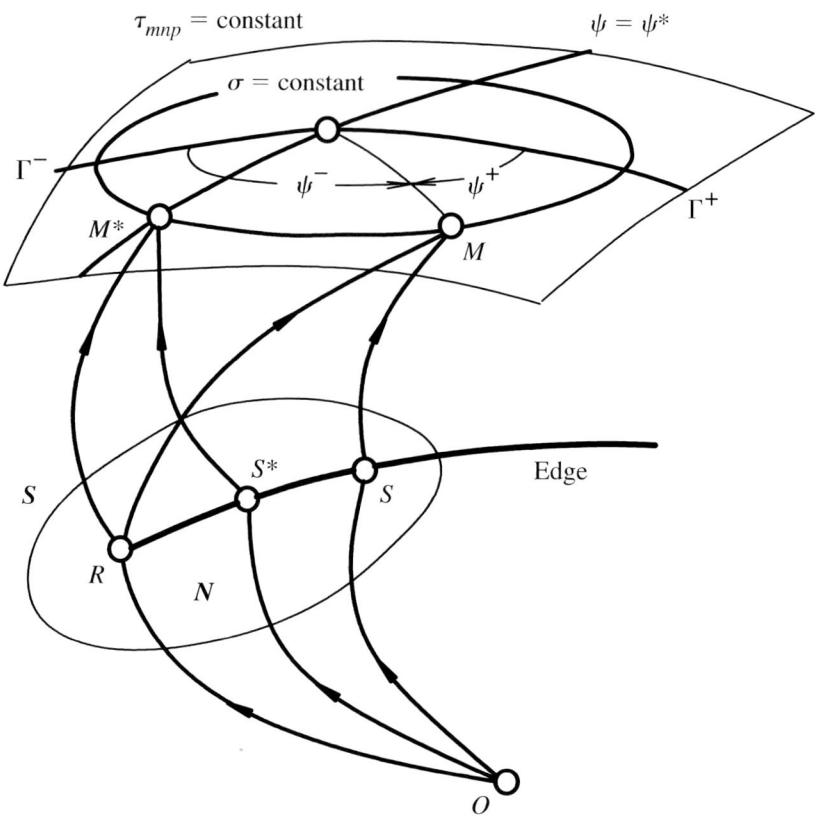

Figure 6. Diffracted rays arising in the neighborhood of an individual tip R. O is the source, M is the observation point, and M^* is its projection on the primary shadow boundary. The maximum size of the region N, bounded by the closed surface S and containing diffraction points S, S^*, and R, can be chosen arbitrarily small when coordinate σ tends to zero (Klem-Musatov, 1994).

approximate expression for the reflected/transmitted wave eikonal in the neighborhood of the central ray by changing index mn to index m in equations 33 and 39:

$$\tau_m = \tau_{mnp} + \frac{\sigma^2 a_m\left(\psi^{\pm}\right)}{2}, \qquad (42)$$

$$\tau_m = \tau_{mnp} - \left(\tau_{mnp} - \tau_m^*\right)A_m, \qquad (43)$$

$$A_m = \frac{a_m\left(\psi^{\pm}\right)}{a_m\left(\psi^*\right)}, \qquad (44)$$

$$a_m\left(\psi^{\pm}\right) = \left(\frac{\partial^2 \tau_m}{\partial \sigma^2}\right)_{\sigma=0}, \tag{45}$$

and

$$\tau_m^* = \tau_m\left(\tau_{mnp}, \psi^*, \sigma\right), \tag{46}$$

where ψ^* is the coordinate of the primary shadow boundary in accordance with equation 37.

Because the fronts of reflected/transmitted and edge waves are tangent at the primary shadow boundary, this equality holds true:

$$\tau_{mn}^* = \tau_m^*. \tag{47}$$

The principle of locality

We can find explicit expressions for functions 40 and 44, using the following considerations. Take, for example, expression 39. Solving for A_{mn}, we obtain

$$A_{mn} = \frac{\tau_{mnp} - \tau_{mn}}{\tau_{mnp} - \tau_{mn}^*}, \tag{48}$$

where eikonals τ_{mn} and τ_{mn}^* depend on σ. On the strength of definitions 34 and 40, however, the same quantity can be represented in the form

$$A_{mn} = \frac{\left[\dfrac{\partial^2 \tau_{mn}\left(\tau_{mnp}, \psi^{\pm}, \sigma\right)}{\partial \sigma^2}\right]_{\sigma=0}}{\left[\dfrac{\partial^2 \tau_{mn}\left(\tau_{mnp}, \psi^*, \sigma\right)}{\partial \sigma^2}\right]_{\sigma=0}}, \tag{49}$$

which does not depend on the coordinate σ.

From expressions 48 and 49, it follows that

$$A_{mn} = \lim_{\sigma \to 0} \frac{\tau_{mnp} - \tau_{mn}}{\tau_{mnp} - \tau_{mn}^*}. \tag{50}$$

Applying the same considerations to equation 44, we obtain

$$A_m = \lim_{\sigma \to 0} \frac{\tau_{mnp} - \tau_m}{\tau_{mnp} - \tau_m^*}. \tag{51}$$

Consider one important property of expressions 50 and 51. We denote a tip by R and edge points corresponding to eikonals τ_{mn} and τ_{mn}^* by S and S^* (see Figure 6). Let N be a region of space bounded by the closed surface S. Let this region include all three points, R, S, and S^*, and the part of the reflected/transmitted ray nearest to them. Surface S always can be chosen so that traveltimes along the rays OS, OS^*, OR, and OM from the real or imaginary source O to the surface S are all equal to one another. Traveltimes from surface S along the rays SM, S^*M^*, RM, and OM are also equal. Then all differences of traveltimes in equations 50 and 51 outside region N equal zero, and the corresponding quantities depend only on the differences of eikonals within region N.

As $\sigma \to 0$, points S and S^* tend to point R. The distance between the reflected/transmitted ray and point R also tends to zero. Therefore, as $\sigma \to 0$, we can choose the region N as a small neighborhood about the point R. Within such a small region, it is possible to neglect the inhomogeneity of the medium and the curvatures of wavefronts and interfaces. Then the value of quantity 50 or 51 would be the same in all situations with the same description in a small neighborhood of the tip. Because quantities 50 and 51 depend only on conditions in a neighborhood of the tip, this fact can be expressed as the following principle of locality. It is sufficient to find them in the simplest situation (e.g., diffraction of the plane wave on the rectangular edge in a homogeneous medium) and then use the expressions obtained in general situations (inhomogeneous media with curvilinear interfaces and curvilinear wavefronts).

Using this principle, we can determine quantity 51 easily. Indeed, in the case of a plane reflected/transmitted wave, the value of the eikonal τ_m does not depend on the value of coordinate ψ^*. Therefore,

$$\tau_m\left(\tau_{mnp}, \psi^\pm, \sigma\right) = \tau_m\left(\tau_{mnp}, \psi, \sigma\right), \tag{52}$$

where ψ is any value in interval $|\psi| \leq \pi$. Taking $\psi = \psi^*$, we have

$$\tau_m^* = \tau_m. \tag{53}$$

Substituting this expression into equation 51, we obtain

$$A_m = 1. \tag{54}$$

From equations 47 and 53, it follows that

$$\tau_{mn}^* = \tau_m\left(\tau_{mnp}, \psi^\pm, \sigma\right), \tag{55}$$

where $\psi^\pm$ is any value from interval $|\psi^\pm| \leq \pi$.

Substituting expression 55 into 39, we obtain the following representation of the edge-wave eikonal:

$$\tau_{mn} = \tau_{mnp} - \left(\tau_{mnp} - \tau_m\right)A_{mn}, \tag{56}$$

where all eikonals are functions of the arbitrary point M. Next we will use the principle of locality to find the quantity A_{mn} in this expression. To show the independence of this quantity on the curvature of the reflected/transmitted wavefront, we will increase the problem's complexity by replacing the plane wave with a spherical one.

The case of the spherical wave

Consider the case of a spherical wave striking a rectangular edge in a homogeneous medium. To do this, first we will derive the approximate expression of the edge-wave eikonal in the neighborhood of the central ray. This can be done by approximating the edge wavefront by a surface of the second order.

Denote by ρ the distance between the current point M and the tip R, and denote by σ the angle between the central ray OT and vector RM, where point O corresponds to the real or imaginary source of the spherical wave (Figure 7). Denote the angle between a pair of planes Π and Π_1 containing the central ray by $\psi^{\pm}$. Let plane Π contain point M, and let the position of plane Π_1 be fixed. Then the position of point M can be determined by values of ρ, $\psi^{\pm}$, and σ, now spatial coordinates.

Let $\ell = \ell(\psi^{\pm})$ denote the radius of curvature of the normal cross section of the edge wavefront $\tau_{mn} = \tau_{mn}(\rho, \psi^{\pm}, \sigma)$ in plane Π at central ray $\sigma = 0$. In the neighborhood of the central ray, the edge wavefront can be approximated by the surface formed by lines of equal curvature

$$\ell\left(\rho, \psi^{\pm}, \sigma\right) = \text{constant}, \tag{57}$$

where the value of ℓ (with $\psi^{\pm}$ and ρ fixed) does not depend on σ. Then the edge-wave eikonal can be represented as follows:

$$\tau_{mn} = \frac{\ell\left(\rho, \psi^{\pm}, \sigma\right)}{c_m} + \delta\left(\psi^{\pm}\right) \quad \text{as} \quad \sigma \to 0, \tag{58}$$

where $\delta(\psi^{\pm})$ is the eikonal value (which does not depend on σ) and c_m is the propagation velocity.

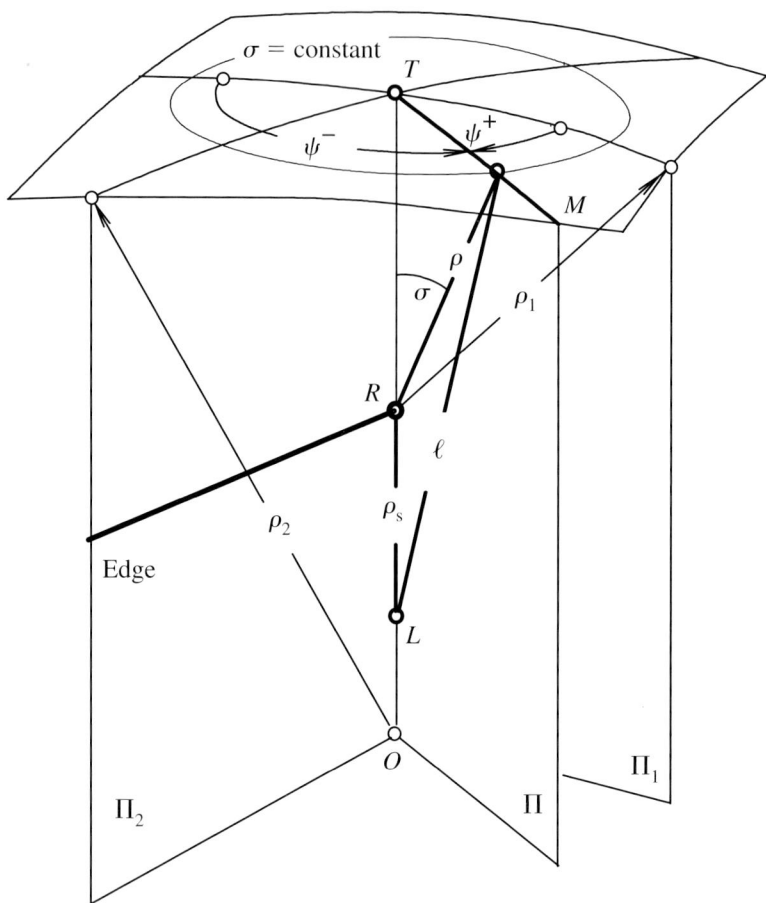

Figure 7. Approximation of the edge wavefront by lines of equal curvature. O is the source, M is the observation point, and R is the tip. The planes Π_1 and Π_2 are normal main cross sections of the edge wavefront. Point L is the center of curvature of the edge wavefront by the plane Π containing central ray OT.

The radius of curvature of the wavefront in equation 58 can be expressed through spatial coordinates by using the cosine theorem as follows:

$$\ell = \sqrt{\rho^2 + \rho_s^2 + 2\rho\rho_s \cos\sigma},\qquad(59)$$

where $\rho_s = \rho_s\left(\psi^\pm\right)$ is the distance between tip R and the center of curvature $\ell = 0$.

Express ρ_s in terms of coordinates ρ and $\psi^\pm$. Let Π_k and ρ_k with $k = 1, 2$ denote planes of normal main cross sections of the edge wavefront and principal radii of curvature of these cross sections at the central ray, i.e., at point T in Figure 7. According to the kinematic law of edge diffraction, the edge wavefront is a surface of rotation around the edge. Differential geometry states that one of the planes of the main cross section of such a surface must contain the axis of rotation, i.e., the edge.

Let it be plane Π_2. In the case under consideration, this plane coincides with the primary shadow boundary. Because the main planes are mutually orthogonal, plane Π_1 touches the secondary shadow boundary at the central ray. It is clear that quantity $\psi^\pm$ represents the ray coordinate of the tip wave. The radius of curvature of the edge wavefront's normal cross section by plane Π (forming the angle $\psi^\pm$ with plane Π_1) can be expressed through the main radii by Euler's formula:

$$\frac{1}{\ell} = \frac{\cos^2 \psi^\pm}{\rho_1} + \frac{\sin^2 \psi^\pm}{\rho_2}. \tag{60}$$

Use this formula to find quantity ρ_s from equation 59. When $\sigma = 0$, the following relationships hold true:

$$\ell = \rho + \rho_s, \quad \rho_1 = \rho, \quad \rho_2 = \rho + \rho^*, \quad \rho^* = (\rho_s)_{|\psi^\pm|=\pi/2}. \tag{61}$$

Substituting equations 61 into 60 and solving for ρ_s, we obtain

$$\rho_s = \frac{\rho \rho^* \sin^2 \psi^\pm}{\rho + \rho^* \cos^2 \psi^\pm}. \tag{62}$$

Expression 58 (where ℓ and ρ_s are determined by equations 59 and 62, respectively) approximates the edge wavefront in the neighborhood of the central ray. Three quantities $(\rho, \psi^\pm, \sigma)$ in these expressions can be regarded as the ray coordinates of the tip wave. Indeed, the tip-wave eikonal has the form

$$\tau_{mnp} = \frac{\rho + OR}{c_m},$$

where distance OR and wave velocity c_m are constants. As mentioned above, $\psi^\pm$ with $\sigma \to 0$ coincides with the corresponding ray coordinate of the tip wave. Choice of σ also corresponds to the definition of the tip-wave coordinate. Therefore, we can use the expression obtained to find quantity 40.

Substituting equations 59 and 62 into 58 and differentiating the result of the substitution, we obtain

$$\left(\frac{\partial^2 \tau_{mn}}{\partial \sigma^2}\right)_{\sigma=0} = -\frac{\rho \rho^* \sin^2 \psi^\pm}{c_m(\rho + \rho^*)}. \tag{63}$$

Substituting this expression into 40, we have

$$A_{mn} = \sin^2 \psi^\pm. \tag{64}$$

Notice that this quantity does not depend on the curvature of the incident spherical wave, i.e., it has the same form as in the case of the incident plane wave. This is the simplest demonstration of the principle of locality.

Variables ζ_{mnp} and ρ_{mnp}

Introduce new variables convenient for the description of tip diffraction. Substituting expression 64 into 56, we obtain

$$\tau_{mn} = \tau_{mnp} - \left(\tau_{mnp} - \tau_m\right)\sin^2 \psi^\pm. \tag{65}$$

Solving this expression for $\psi^\pm$, we obtain

$$\psi^\pm = \arcsin \sqrt{\frac{\tau_{mnp} - \tau_{mn}}{\tau_{mnp} - \tau_m}}. \tag{66}$$

This expression can be used to find the absolute value of coordinate $\psi^\pm$ through the eikonal values with $\psi^\pm \leq \pi/2$. Indeed, we can consider τ_{mn} here as a function of this coordinate

$$\tau_{mn} = \tau_{mn}\left(\tau_{mnp}, \psi^\pm, \sigma\right) \quad \text{with} \quad \tau_{mnp} = \text{constant}, \quad \sigma = \text{constant}, \tag{67}$$

changing from $\tau_{mn} = \tau_{mnp}$ for $\psi^\pm = 0$ or $|\psi^\pm| = \pi$ to $\tau_{mn} = \tau_m$ for $|\psi^\pm| = \pi/2$. Then equation 66 changes from $\psi^\pm = 0$ to $\psi^\pm = \pi/2$ within interval $0 \leq \psi^\pm \leq \pi/2$. One can say that expression 67 determines the principal positive value of the trigonometric function under consideration. We introduce special notation for this principal value:

$$\zeta_{mnp} = \arcsin \sqrt{\frac{\tau_{mnp} - \tau_{mn}}{\tau_{mnp} - \tau_m}}. \tag{68}$$

Any value of $\psi^{\pm}$ can be expressed through this value according to equation 10 as follows:

$$\psi^+ = s_{mnp}\zeta_{mnp}, \quad \psi^- = -s_{mnp}(\zeta_{mnp} - \pi) \quad \text{for} \quad |\psi^+| < \pi/2,$$
$$\psi^+ = s_{mnp}(\zeta_{mnp} - \pi), \quad \psi^- = -s_{mnp}\zeta_{mnp} \quad \text{for} \quad |\psi^-| < \pi/2, \quad (69)$$

or in the more compact form

$$\psi^{\pm} = \pm s_{mnp}\zeta_{mnp}, \quad \psi^{\mp} = \mp s_{mnp}(\zeta_{mnp} - \pi) \quad \text{for} \quad |\psi^{\pm}| < \pi/2. \quad (70)$$

We also introduce the quantity

$$\rho_{mnp} = \sqrt{\frac{2\omega(\tau_{mnp} - \tau_m)}{\pi}}. \quad (71)$$

In the neighborhood of the central ray, τ_m is a function of two coordinates (see section titled "The principle of locality" in this chapter),

$$\tau_m = \tau_m(\tau_{mnp}, \sigma), \quad (72)$$

resulting in 71 also being a function of two coordinates, τ_{mnp} and σ, so that

$$\rho_{mnp} = \rho_{mnp}(\tau_{mnp}, \sigma). \quad (73)$$

One can consider the three quantities $(\tau_{mnp}, \zeta_{mnp}, \rho_{mnp})$ as coordinates of an arbitrary point in space, within an individual domain of continuity, determined in the section titled "Domains of continuity" in this chapter. Using expressions 70 and 71, we can represent the edge-wave eikonal 65 as a function of the new variables in the form

$$\tau_{mn} = \tau_{mnp} - \frac{\pi\rho_{mnp}^2}{2\omega}\sin^2\zeta_{mnp}. \quad (74)$$

The edge-wave amplitude

Consider the approximation of the edge-wave amplitude in the neighborhood of the secondary shadow boundary. According to equation 57 of Chapter 3, this amplitude is determined by the expression

$$\Phi_{mn} = s_{mn}\Phi_m W(w_{mn}), \quad w_{mn} = \sqrt{\frac{2\omega(\tau_{mn} - \tau_m)}{\pi}}, \quad (75)$$

where the unit function s_{mn} is determined by equation 58 of Chapter 3.

When finding the edge wave, we approximated the reflected/transmitted wave amplitude by its value at the primary shadow boundary 13 of Chapter 3 because that amplitude was regarded as a slowly changing function of a point in space. However, now we cannot assume that the edge-wave amplitude will change in a similar manner without preliminary analysis. This is because the neighborhood of the central ray is a region where edge-wave amplitude varies rapidly. Therefore, it would seem that a similar approximation is not permissible. At the same time, it follows from the qualitative image of the edge-wave amplitude's 3D distribution that it changes slowly in the direction normal to the secondary shadow boundary. We use these considerations to gain a better understanding of further approximations but not to justify these approximations (necessary justification will be given later).

Recall that the secondary shadow boundary is formed by a cone of diffracted rays emitted from the tip. In Chapter 2, in the section titled "Transverse diffusion," we mentioned that during transverse diffusion, no energy is exchanged between the cones of diffracted rays (there are no derivatives in directions normal to the cone surface in the equation of transverse diffusion). Therefore, edge-wave amplitude must change slowly in the direction normal to the cone surface, but then the intensity of the edge wave as a function of coordinate $\psi^{\pm}$ must have an extremum at the secondary shadow boundary $\psi^{\pm} = 0$.

Qualitative interpretation of the 3D change in argument w_{mn} of function $W(w_{mn})$ in equation 75 above leads to the same conclusion. Although varying the value of $\psi^{\pm}$ in the neighborhood of the secondary shadow boundary with small values of σ, the difference $\tau_{mn} - \tau_m$ is practically constant. That is because the tangent to the coordinate $\psi^{\pm}$-line is almost parallel to the plane, which touches the primary shadow boundary $\tau_{mn} = \tau_m$ at the central ray in this case. Therefore, quantity W in equation 75 is practically constant also, and the edge-wave amplitude must have an extremum at the secondary shadow boundary. Thus it seems that we can approximate the edge-wave amplitude by its value at the secondary shadow boundary. We will justify this conclusion later by quantitative analysis.

To make such an approximation, expand the edge-wave amplitude in a power series in coordinate $\psi^{\pm}$ in the neighborhood of the secondary shadow boundary and estimate the relative contributions of the individual items. For this analysis only, it is sufficient to use the linear part of such an expansion:

$$\Phi_{mn}\left(\tau_{mnp}, \psi^{\pm}, \sigma\right) = \left[\Phi_{mn}\left(\tau_{mnp}, \psi^{\pm}, \sigma\right)\right]_{\psi^{\pm}=0} + \varepsilon, \qquad (76)$$

and

$$\varepsilon = \Phi'_{mn} \psi^{\pm}, \quad \Phi'_{mn} = \left(\frac{\partial \Phi_{mn}}{\partial \psi^{\pm}} \right)_{\psi^{\pm}=0}. \tag{77}$$

Next we will show that the second term in equation 76 has the asymptotic estimate

$$\varepsilon \sim O\left(\frac{1}{\sqrt{\omega}} \right), \tag{78}$$

which allows us to neglect this term as $\omega \to \infty$ so that the following approximation can be used:

$$\Phi_{mn}\left(\tau_{mnp}, \psi^{\pm}, \sigma \right) = \left[\Phi_{mn}\left(\tau_{mnp}, \psi^{\pm}, \sigma \right) \right]_{\psi^{\pm}=0}. \tag{79}$$

Derivation of the asymptotic estimate

Perform the formal derivation in equation 77, taking into account the following considerations: First, the unit function s_{mn} is constant in the neighborhood of the secondary shadow boundary. Second, the reflected/transmitted-wave amplitude is a slowly changing function in the neighborhood of the secondary shadow boundary. This allows us to use the approximation

$$\Phi_m\left(\tau_{mnp}, \psi^{\pm}, \sigma \right) = \left[\Phi_m\left(\tau_{mnp}, \psi^{\pm}, \sigma \right) \right]_{\psi^{\pm}=0} = \Phi_m\left(\tau_{mnp}, 0, 0 \right). \tag{80}$$

Taking all this into account and substituting expression 75 into 77, we write

$$\Phi'_{mn} = s_{mn} \Phi_m \left(W' w' \right)_{\psi^{\pm}=0}, \quad W' = \frac{dW(w_{mn})}{dw_{mn}}, \quad w' = \frac{\partial w_{mn}}{\partial \psi^{\pm}} \tag{81}$$

where, according to equation 110 of Chapter 2,

$$W' = -\frac{1}{\sqrt{2}} \exp\left(-\frac{i\pi}{4} \right) - i\pi w_{mn} W(w_{mn}). \tag{82}$$

Find the derivative w'. Let ds be an element along the arc length of the $\psi^{\pm}$-coordinate line, i.e., the line of intersection of surfaces $\tau_{mnp} = $ constant and $\sigma = $ constant. Then

$$\frac{ds}{d\psi^{\pm}} = r_s, \tag{83}$$

where r_s is the radius of the curvature of the element of the arc length ds, and the required derivative can be represented as

$$w' = \frac{dw_{mn}}{ds}\frac{ds}{d\psi^{\pm}} = r_s\frac{dw_{mn}}{ds}. \tag{84}$$

By considering the resulting derivative as directional,

$$\frac{dw_{mn}}{ds} = \left(\nabla w_{mn}\cdot\mathbf{i}^{\pm}\right), \tag{85}$$

where $\mathbf{i}^{\pm}$ is a unit vector of derivation, i.e., the unit vector of the tangent to $\psi^{\pm}$-coordinate line, we obtain the following form of expression 84:

$$w' = r_s\left(\nabla w_{mn}\cdot\mathbf{i}^{\pm}\right). \tag{86}$$

Develop the scalar product in equation 86. Using expression for w_{mn} from 75, we obtain

$$\nabla w_{mn} = \nabla\sqrt{\frac{2\omega\left(\tau_{mn} - \tau_m\right)}{\pi}} = \varphi\left(\nabla\tau_{mn} - \nabla\tau_m\right) \tag{87}$$

and

$$\varphi = \sqrt{\frac{\omega}{2\pi\left(\tau_{mn} - \tau_m\right)}}. \tag{88}$$

Substituting expression 87 into 86, we obtain

$$w' = r_s\varphi\left[\left(\nabla\tau_{mn}\cdot\mathbf{i}^{\pm}\right) - \left(\nabla\tau_m\cdot\mathbf{i}^{\pm}\right)\right]$$
$$= r_s\varphi\left[\left|\nabla\tau_{mn}\right|\cos\left(\nabla\tau_{mn}, \mathbf{i}^{\pm}\right) - \left|\nabla\tau_m\right|\cos\left(\nabla\tau_m, \mathbf{i}^{\pm}\right)\right] = \frac{r_s\varphi B}{c_m},$$
$$B = \cos\left(\nabla\tau_{mn}, \mathbf{i}^{\pm}\right) - \cos\left(\nabla\tau_m, \mathbf{i}^{\pm}\right). \tag{89}$$

While deriving this expression, we used the equalities $\left|\nabla\tau_{mn}\right| = \left|\nabla\tau_{m}\right| = 1/c_m$ where c_m is the wave velocity.

We represent the quantities in equation 89 as explicit functions of coordinate σ, using Taylor-series power expansions in the neighborhood of the central ray,

$$r_s = \sum_{k=0}^{\infty} q_k \sigma^k, \quad q_k = \frac{1}{k}\left(\frac{\partial^k r_s}{\partial \sigma^k}\right)_{\sigma=0}. \tag{90}$$

Because $r_s \to 0$ as $\sigma \to 0$, we have $q_0 = 0$. Neglecting the second and higher powers of this expansion, we obtain

$$r_s = q_1 \sigma. \tag{91}$$

At the secondary shadow boundary, the difference of the eikonals in equation 88 can be represented by equation 14 of Chapter 3 as

$$\tau_{mn} - \tau_m = -a_2 \alpha^2 \quad \text{for} \quad \psi^{\pm} = 0, \tag{92}$$

where a_2 is determined by equation 15 of Chapter 3. Let, for example, coordinate σ have the meaning of the arc length on the surface $\tau_{mnp} = \text{constant}$. Then at the secondary shadow boundary, we have

$$\sigma = r_\sigma \alpha \quad \text{for} \quad \psi^{\pm} = 0, \tag{93}$$

where r_σ is the radius of curvature of the arc length of the element σ. Using this expression, rewrite formula 92 as

$$\tau_{mn} - \tau_m = -\frac{a_2 \sigma^2}{r_\sigma^2} \quad \text{for} \quad \psi^{\pm} = 0. \tag{94}$$

Substituting this expression into 88, we obtain

$$\varphi = \frac{r_\sigma}{\sigma}\sqrt{-\frac{\omega}{2\pi a_2}} \quad \text{for} \quad \psi^{\pm} = 0. \tag{95}$$

Now we represent quantity B in equation 89 by the power series

$$B = \sum_{k=0}^{\infty} b_k \sigma^k, \quad b_k = \frac{1}{k}\left(\frac{\partial^k B}{\partial \sigma^k}\right)_{\sigma=0}. \tag{96}$$

Because vectors $\nabla \tau_{mn}$ and $\nabla \tau_m$ are orthogonal to the vector $\mathbf{i}^{\pm}$ with $\sigma = 0$, we have $b_0 = 0$. Neglecting the second and higher powers in the series, we obtain

$$B = b_1 \sigma. \tag{97}$$

By substituting expressions 91, 95, and 97 into 89, we obtain

$$(w')_{\psi^{\pm}=0} = \frac{b\sigma}{\sqrt{\omega}}, \quad b = \left(\frac{q_1 b_1 r_\sigma}{c_m \sqrt{-2\pi a_2}} \right)_{\psi^{\pm}=0}. \tag{98}$$

We find the asymptotic estimation of quantity 77 by substituting expressions 81 and 98 into 77:

$$\varepsilon = C \sigma \psi^{\pm} \sqrt{\omega}, \quad C = s_{mn} \Phi_m b (W')_{\psi^{\pm}=0}. \tag{99}$$

Quantity C is bounded in the neighborhood of the central ray. Quantity σ has the asymptotic estimate 36, and quantity $\psi^{\pm}$ has an estimate of boundary-layer type 153 of Chapter 2 so that

$$\sigma \sim O\left(\frac{1}{\sqrt{\omega}} \right), \quad \psi^{\pm} \sim O\left(\frac{1}{\sqrt{\omega}} \right) \quad \text{as} \quad \omega \to \infty. \tag{100}$$

Finally,

$$\sigma \psi^{\pm} \sqrt{\omega} \sim O\left(\frac{1}{\sqrt{\omega}} \right) \quad \text{as} \quad \omega \to \infty. \tag{101}$$

This proves the asymptotic estimate in equation 78.

The edge wave

Using equations 74 and 79, we can represent the edge wave in the neighborhood of the secondary shadow boundary as

$$f_{mn} = \Phi_{mn} \exp(i\omega\tau_{mn}) = A \exp(-i\nu \sin^2 \zeta_{mnp}), \tag{102}$$

$$A = \left[\Phi_{mn} (\tau_{mnp}, \psi^{\pm}, \sigma) \right]_{\psi^{\pm}=0} \exp(i\omega\tau_{mnp}), \tag{103}$$

and

$$\nu = \frac{\pi \rho_{mnp}^2}{2}. \tag{104}$$

Quantity 102 is a function of the real variable ζ_{mnp} with the values of τ_{mnp} and σ fixed. Consider quantity 102 as a set of values of a function of the complex variable on the real axis. Therefore, we introduce the complex variable $\alpha = x + iy$ instead of the real variable ζ_{mnp} and write equation 102 as

$$f_{mn}(\alpha) = A \exp\left(-i\nu \sin^2 \alpha\right), \tag{105}$$

where A and ν do not depend on α. This expression gives an analytic function of the complex variable α at all finite points of the complex plane.

To analyze the behavior of this function as $\alpha \to \infty$, separate the real and imaginary parts of the exponent

$$f_{mn}(\alpha) = A \exp\left(R + iI\right), \tag{106}$$

$$R = (\nu/2)\sin 2x \cdot \sinh 2y, \quad I = \nu\left(\cos^2 x \cdot \sinh^2 y - \sin^2 x \cdot \cosh^2 y\right), \tag{107}$$

and consider the modulus of this function

$$\left|f_{mn}(\alpha)\right| = \left|A\right| \exp R. \tag{108}$$

One can see that the behavior of quantity 108, as its argument becomes infinite, is given by the expressions

$$\left|f_{mn}(\alpha)\right| \to 0 \quad \text{as} \quad \alpha \to \infty, \quad \text{if} \quad R < 0, \tag{109}$$

and

$$\left|f_{mn}(\alpha)\right| \to \infty \quad \text{as} \quad \alpha \to \infty, \quad \text{if} \quad R > 0. \tag{110}$$

To interpret these conditions geometrically, consider the expression

$$\text{sign}\,(R) = \text{sign}\,(\nu) \cdot \text{sign}\,(\sin 2x \cdot \sinh 2y). \tag{111}$$

From equations 104 and 71, it follows that

$$\text{sign}\,(\nu) = \text{sign}\,(\tau_{mnp} - \tau_m). \tag{112}$$

By taking into account this expression and the relationships

$$\text{sign}\left(\sin 2x \cdot \sinh 2y\right) = \text{sign}\left(\sin 2x\right)\text{sign}\left(y\right) \tag{113}$$

and

$$\begin{aligned}\sin 2x > 0 &\quad\text{for}\quad \pi k < x < \pi k + \pi/2, \\ \sin 2x < 0 &\quad\text{for}\quad \pi k - \pi/2 < x < \pi k,\end{aligned} \tag{114}$$

for $k = 0, \pm 1, \pm 2, \ldots$, rewrite equation 111 as

$$\text{sign}\left(R\right) = \text{sign}\left(\tau_{mnp} - \tau_m\right) \cdot \text{sign}\left(y\right) \cdot s\left(x\right) \tag{115}$$

and

$$\begin{aligned}s(x) = +1 &\quad\text{for}\quad \pi k < x < \pi k + \pi/2, \\ s(x) = -1 &\quad\text{for}\quad \pi k - \pi/2 < x < \pi k,\end{aligned} \tag{116}$$

for $k = 0, \pm 1, \pm 2, \ldots$.

Using expression 115, we can determine the domains of the complex plane α, where function 106 is analytic as $\alpha \to \infty$, i.e., where it satisfies condition 109, as follows:

$$\text{sign}\left(\tau_{mnp} - \tau_m\right) \cdot \text{sign}\left(y\right) \cdot s\left(x\right) < 0. \tag{117}$$

These domains are illustrated in Figure 8 with hatching.

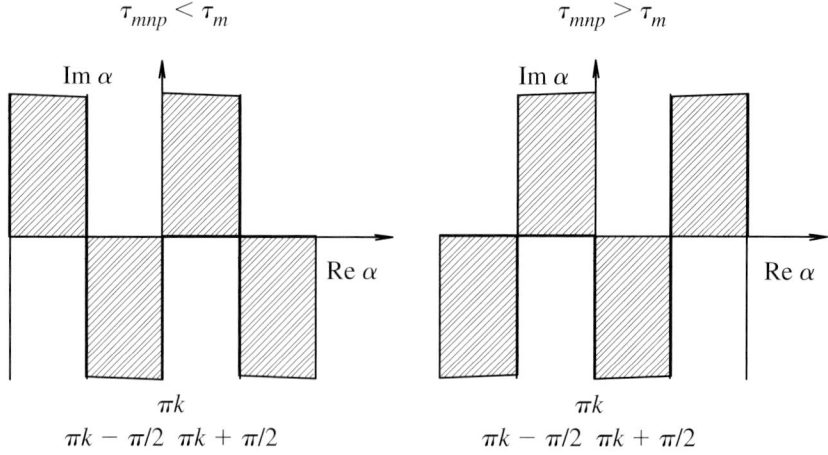

Figure 8. Domains of analyticity (shaded areas) of the edge wave as a function of complex variable α (Klem-Musatov, 1994).

Thus, in the neighborhood of the secondary shadow boundary, we can regard edge wave 102 as a function of complex variable 105 of Chapter 4, tending to zero at infinity in the regions 117 above.

Smoothing the discontinuity at a secondary shadow boundary

The condition of periodicity

We seek components $f^{\pm}(\psi^{\pm})$ of tip wave 5, using the condition of continuity at secondary shadow boundary 20. Because coordinate $\psi^{\pm}$ varies within finite interval $|\psi^{\pm}| \leq \pi$, we must satisfy the continuity of these components at surface $|\psi^{\pm}| = \pi$. Therefore, we introduce a special representation of these components that guarantee this condition, based on the obvious property of periodicity of the fields under consideration over coordinate $\psi^{\pm}$,

$$f_{mn}(\psi^{\pm} + 2\pi k) = f_{mn}(\psi^{\pm}) \tag{118}$$

and

$$f^{\pm}(\psi^{\pm} + 2\pi k) = f^{\pm}(\psi^{\pm}), \tag{119}$$

where k is any whole number.

Assume that function $f^{\pm}$ can be represented by the infinite series

$$f^{\pm}(\psi^{\pm}) = \sum_{k=-\infty}^{\infty} F^{\pm}(\psi^{\pm} + 2\pi k), \tag{120}$$

where $F^{\pm}$ is some unknown function. Then $f^{\pm}(\psi^{\pm})$ complies with condition 119 because replacement of $\psi^{\pm}$ by $\psi^{\pm} + 2\pi k$ leads only to new notation for the summation index. Later, we will see that the infinite series exists.

Substituting expression 120 into 20, we write the condition of continuity at the secondary shadow boundary in the form

$$\left[\sum_{k=-\infty}^{\infty} F^{\pm}(\psi^{\pm} + 2\pi k) + f_{mn}(\psi^{\pm}) \right]_{\psi^{\pm}=0_-} = \left[\sum_{k=-\infty}^{\infty} F^{\pm}(\psi^{\pm} + 2\pi k) \right]_{\psi^{\pm}=0_+}. \tag{121}$$

Assume that the function $F^\pm$ satisfies the following conditions of continuity at the secondary shadow boundary:

$$\left[F^\pm(\psi^\pm) + f_{mn}(\psi^\pm)\right]_{\psi^\pm=0_-} = \left[F^\pm(\psi^\pm)\right]_{\psi^\pm=0_+}, \tag{122}$$

$$\left[\sum_{k=-\infty}^{\infty} F^\pm(\psi^\pm + 2\pi k)\right]_{\psi^\pm=0_-} = \left[\sum_{k=-\infty}^{\infty} F^\pm(\psi^\pm + 2\pi k)\right]_{\psi^\pm=0_+} \quad \text{for } k \neq 0. \tag{123}$$

Then this function also satisfies condition 121.

A nonperiodic function

We introduce the new variable

$$\eta^\pm = \psi^\pm + 2\pi k, \tag{124}$$

where k is any integer. Although $\psi^\pm$ varies within bounded interval $-\pi \leq \psi^\pm \leq \pi$, variable 124 can vary within the infinite interval $(-\infty, \infty)$ in correspondence with the value of k. It becomes zero only if $\psi^\pm = 0$, $k = 0$. Consider all quantities in expressions 118 through 120 as functions of the new variable, varying within the infinite interval $-\infty < \eta^\pm < \infty$:

$$f_{mn}(\psi^\pm) = f_{mn}(\psi^\pm + 2\pi k) = f_{mn}(\eta^\pm), \tag{125}$$

$$f^\pm(\psi^\pm) = f^\pm(\psi^\pm + 2\pi k) = f^\pm(\eta^\pm), \tag{126}$$

and

$$F^\pm(\pi^\pm + 2\pi k) = F^\pm(\eta^\pm). \tag{127}$$

Then f_{mn} and $f^\pm$ are 2π-periodic functions of the new variable. However, function $F^\pm(\eta^\pm)$ is not periodic. Indeed, according to condition 122, it must have a discontinuity at $\eta^\pm = 0$. In addition, according to condition 123, it must be continuous at $\eta^\pm = 2\pi k$ for any $k \neq 0$. It is clear that only a nonperiodic function can comply with both conditions.

Assume that the function $F^\pm(\eta^\pm)$ satisfies

$$\left[F^\pm(\eta^\pm) + f_{mn}(\eta^\pm)\right]_{\eta^\pm=0_-} = \left[F^\pm(\eta^\pm)\right]_{\eta^\pm=0_+} \tag{128}$$

and is continuous when $\eta^\pm \neq 0$. Then conditions 122 and 123 are satisfied automatically, and therefore, the condition at the secondary shadow boundary 121 also is satisfied.

Representation of the tip wave in the form 5, 120, and 127 allows us to write the condition of continuity at the secondary shadow boundary in the form 128. Then it is sufficient for function $F^\pm(\eta^\pm)$ to be continuous at $\eta^\pm \neq 0$ to guarantee continuity of the desired functions at $|\psi^\pm| = \pi$.

The problem of smoothing a discontinuity

Using representation 127, we can reduce finding tip-wave components $f^\pm$ to the problem of finding a function $F^\pm(\eta^\pm)$ that satisfies condition 128. To state the corresponding problem, it is sufficient to repeat the derivations given in Chapter 3, in the section titled "Smoothing the discontinuity at the shadow boundary."

In the section titled "The edge wave" in this chapter, we illustrated that function $f_{mn}(\eta^\pm)$ in condition 128 can be regarded as an analytic function 106 of the complex variable α tending to zero, as $\alpha \to \infty$ in regions 117. By using these properties of the function $f_{mn}(\eta^\pm)$, we can reduce the problem under consideration to that of Sohotsky-Plemelj.

Take a smooth, nonself-intersecting, infinite-oriented contour Γ in the complex plane of α. Let this contour cross the real axis at point $\alpha = 0$ in the direction from the upper semiplane to the bottom semiplane (Figure 9). Let the infinite parts of the contour belong to regions 117,

$$\begin{cases} \text{sign}\,(\text{Im}\,\alpha)\cdot s(\text{Re}\,\alpha) < 0 & \text{for} \quad \tau_{mnp} > \tau_m, \\ \text{sign}\,(\text{Im}\,\alpha)\cdot s(\text{Re}\,\alpha) > 0 & \text{for} \quad \tau_{mnp} < \tau_m, \end{cases} \tag{129}$$

where function $s(x)$ is determined by equations 116. Denote by Ω_- and Ω_+ semi-infinite regions of the complex plane on the right- and left-hand sides of contour Γ, in accordance with the contour's orientation. Using equation 105, introduce the analytic function on Γ,

$$f_{mn}(\alpha) = A\,\exp\left(-i\nu\sin^2\alpha\right) \quad \text{for} \quad \alpha \in \Gamma, \tag{130}$$

where quantities A and ν are determined by equations 103 and 104. This function is Holder continuous on contour Γ (see equation 8 of Chapter 3):

$$\left|f_{mn}(\alpha + \Delta\alpha) - f_{mn}(\alpha)\right| \leq C|\Delta\alpha|^\mu \quad \text{as} \quad |\Delta\alpha| \to 0, \quad 0 < \mu < 1, \tag{131}$$

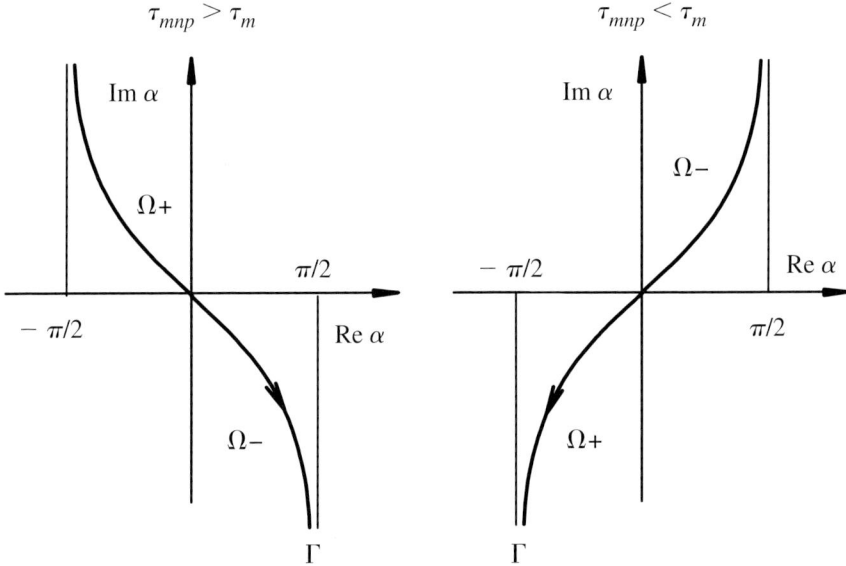

Figure 9. Contour of integration in the domains of analyticity of the edge wave depending on the sign of difference $\tau_{mnp} - \tau_m$ (Klem-Musatov, 1994).

where C is a positive constant, and it vanishes at infinity (see equation 9 of Chapter 3),

$$|f_{mn}(\alpha)| \to 0 \quad \text{as} \quad \alpha \to \infty \quad (\alpha \in \Gamma). \tag{132}$$

Introduce a piecewise-analytic function $F^{\pm}(\alpha)$, satisfying the conditions

$$F^{\pm}_+(\alpha) - F^{\pm}_-(\alpha) = f_{mn}(\alpha) \quad \text{for} \quad \alpha \in \Gamma \tag{133}$$

and

$$|F^{\pm}(\alpha)| \to 0 \quad \text{as} \quad \alpha \to \infty, \tag{134}$$

where $F^{\pm}_+$ and $F^{\pm}_-$ are the boundary values of this function on the left- and right-hand sides of contour Γ, respectively.

Finding this function can be reduced to the Sohotsky-Plemelj problem (see Chapter 3, the section titled "The Sohotsky-Plemelj problem"), the unique solution of which is represented by a Cauchy-type integral (see equation 12 of Chapter 3),

$$F^{\pm}(\alpha) = \frac{1}{2\pi i} \int_{\Gamma} \frac{f_{mn}(\eta)}{\eta - \alpha} d\eta, \tag{135}$$

where α is a complex variable.

Expression 130 on the real axis, Im $\alpha = 0$, coincides with representation of the edge wave in the neighborhood of the secondary shadow boundary 15, 16, and 102. Condition 133 on the real axis, Im $\alpha = 0$, coincides with the condition of continuity at secondary shadow boundary 128.

This means that function 135 smoothes discontinuity of the edge wave at the secondary shadow boundary for real values of variable $\alpha = \eta^{\pm}$. Thus, the desired function is determined by expression 135 with $\alpha = \eta^{\pm}$,

$$F^{\pm}(\eta^{\pm}) = \frac{1}{2\pi i} \int_{\Gamma} \frac{f_{mn}(\alpha)}{\alpha - \eta^{\pm}} d\alpha, \tag{136}$$

where the complex variable η again is denoted as α.

Summation of the series

Returning to the old variable $\psi^{\pm}$ in equation 136, as specified in expression 124, write

$$F^{\pm}(\psi^{\pm} + 2\pi k) = \frac{1}{2\pi i} \int_{\Gamma} \frac{f_{mn}(\alpha)}{\alpha - (\psi^{\pm} + 2\pi k)} d\alpha. \tag{137}$$

Substituting this integral into expression 120 and reversing the order of summation and integration, we obtain

$$f^{\pm}(\psi^{\pm}) = \sum_{k=-\infty}^{\infty} \frac{1}{2\pi i} \int_{\Gamma} \frac{f_{mn}(\alpha)}{\alpha - (\psi^{\pm} + 2\pi k)} d\alpha$$

$$= \frac{1}{2\pi i} \int_{\Gamma} f_{mn}(\alpha) \sum_{k=-\infty}^{\infty} \frac{1}{(\alpha - \psi^{\pm}) - 2\pi k} d\alpha. \tag{138}$$

The sum of the infinite series can be found using decomposition of the cotangent function into partial fractions as

$$\frac{1}{2}\cot\frac{z}{2} = \sum_{k=-\infty}^{\infty} \frac{1}{z - 2\pi k}. \tag{139}$$

Using formula 139, we can rewrite expression 138 as

$$f^{\pm} = \frac{1}{4\pi i} \int_{\Gamma} f_{mn}(\alpha) \cot \frac{\alpha - \psi^{\pm}}{2} \, d\alpha. \tag{140}$$

The smoothing correction as the superposition of solutions of equations of motion

Expression 140 describes a correction to smooth discontinuity of the edge wavefield at the secondary shadow boundary. Here we will show that this correction satisfies the same equation of motion as the initial reflected/transmitted wave in equation 1. To do this, introduce the new variable of integration $z = \alpha - \psi^{\pm}$ and rewrite expression 140 in the form

$$f^{\pm} = \frac{1}{4\pi i} \int_{\Gamma + \psi^{\pm}} f_{mn}(\psi^{\pm} + z) \cot \frac{z}{2} \, dz, \tag{141}$$

where the new contour $\Gamma + \psi^{\pm}$ crosses the real axis at distance $\psi^{\pm}$ relative to contour Γ.

We mentioned in Chapter 3, in the section titled "The Cauchy-type integral as a superposition of solutions of equations of motion," that function f_{mn} satisfies the same equation of motion as the initial reflected/transmitted wave f_m in equation 4 of chapter 3. Therefore, it is sufficient to show that the tip wave obtained satisfies the same equation as the edge wave. Equation 141 above can be regarded as the superposition of functions $f_{mn}(\psi^{\pm})$, depending on the complex parameter z. If function $f_{mn}(\tau_{mnp}, \psi^{\pm}, \sigma)$ satisfies some linear differential equation with coefficients fixed at $\psi^{\pm} = 0$, then superposition of functions $\cot(z/2) f_{mn}(\tau_{mnp}, \psi^{\pm} + z, \sigma)$ over parameter z (on the strength of the linearity of this operation) satisfies the same equation. Formal dependence of the position of the contour of integration $\Gamma + \psi^{\pm}$ on variable $\psi^{\pm}$ does not affect the linearity of the operation because the contour of integration can be deformed without changing the integral (a property of the integral of an analytic function). This proves the above statement.

If field f_{mn} satisfies some linear differential equation within the domain of continuity $-\pi/2 < \psi^{\pm} < 0$, then expression 141 satisfies the same equation separately within each region $-\pi/2 < \psi^{\pm} < 0$ and $0 < \psi^{\pm} < \pi$ but not in their union. The superposition of fields $f_{mn} + f^{\pm}$ satisfies the same equation in the space cut along the primary shadow boundary $\psi^{\pm} = -\pi/2$. A superposition of type 1, including reflected/transmitted, edge, and tip waves, satisfies the same equation in the whole neighborhood of the central ray.

The tip wave

Substituting expressions 105 and 103 into 140 yields

$$f^{\pm} = \Psi^{\pm}\exp\left(i\omega\tau_{mnp}\right)\left[\Phi_{mn}\left(\tau_{mnp}, \psi^{\pm}, \sigma\right)\right]_{\psi^{\pm}=0}, \tag{142}$$

and

$$\psi^{\pm} = \frac{1}{4\pi i}\int_{\Gamma}\exp\left(-i\nu\,\sin^2\alpha\right)\cot\frac{\alpha - \psi^{\pm}}{2}d\alpha, \tag{143}$$

where ν is determined by equation 104.

In accordance with equation 13, we have

$$\left[\Phi_{mn}\left(\tau_{mnp}, \psi^{\pm}, \sigma\right)\right]_{\psi^{\pm}=0} = \pm\Phi_m\left[W(w_{mn})\right]_{\psi^{\pm}=0}, \tag{144}$$

where w_{mn} is determined in equations 75.

Because eikonals of the edge and tip waves coincide ($\tau_{mn} = \tau_{mnp}$) at the secondary shadow boundary (where $\psi^{\pm} = 0$), the following relationship holds true:

$$\left(w_{mn}\right)_{\psi^{\pm}=0} = \left[\sqrt{\frac{2\omega\left(\tau_{mn} - \tau_m\right)}{\pi}}\right]_{\psi^{\pm}=0} = \sqrt{\frac{2\omega\left(\tau_{mnp} - \tau_m\right)}{\pi}}, \tag{145}$$

or taking into account equation 71,

$$\left(w_{mn}\right)_{\psi^{\pm}=0} = \rho_{mnp}. \tag{146}$$

Substitute expressions 144 and 146 into 142 to obtain

$$f^{\pm} = \pm \Phi_m W(\rho_{mnp}) \psi^{\pm} \exp(i\omega\tau_{mnp}). \tag{147}$$

Substituting expression 147 into 5 yields

$$f_{mnp} = \Phi_m W(\rho_{mnp}) \psi \exp(i\omega\tau_{mnp}), \tag{148}$$

where

$$\psi = \psi^{+} - \psi^{-}. \tag{149}$$

Function $\psi(\rho, \zeta)$

Substitute expression 143 into 149 and write the result as follows:

$$\psi = \frac{1}{4\pi i} \int_{\Gamma} P(\alpha) \exp(-i\nu \sin^2 \alpha) \, d\alpha, \tag{150}$$

where

$$P(\alpha) = \cot\frac{\alpha - \psi^{+}}{2} - \cot\frac{\alpha - \psi^{-}}{2} = \frac{1}{\Delta} \sin\frac{\psi^{+} - \psi^{-}}{2}, \tag{151}$$

and

$$\Delta = \sin\frac{\alpha - \psi^{+}}{2} \sin\frac{\alpha - \psi^{-}}{2}. \tag{152}$$

Transform the integrand to a more convenient form, using expressions for quantities $\psi^{\pm}$ from 70, and write

$$\sin\frac{\psi^{+} - \psi^{-}}{2} = \sin\left[s_{mnp}\left(\zeta_{mnp} - \frac{\pi}{2}\right)\right] = -s_{mnp} \cos\zeta_{mnp}$$

$$\text{for} \quad |\psi^{\pm}| < \frac{\pi}{2}, \tag{153}$$

$$\Delta = \sin\frac{\alpha - \psi^{\pm}}{2} \sin\frac{\alpha - \psi^{\mp}}{2} = \sin\frac{\alpha - s\zeta_{mnp}}{2} \sin\frac{\alpha + s(\zeta_{mnp} - \pi)}{2}$$

$$\text{for} \quad |\psi^{\pm}| < \frac{\pi}{2}, \tag{154}$$

where

$$s = \pm s_{mnp} \quad \text{for} \quad |\psi^{\pm}| < \frac{\pi}{2}. \tag{155}$$

It is easy to rewrite expression 154 as

$$\Delta = -s \sin \frac{\alpha - s\zeta_{mnp}}{2} \cos \frac{\alpha + s\zeta_{mnp}}{2}. \tag{156}$$

Using the equality

$$\sin \beta - \sin \gamma = 2 \sin \frac{\beta - \gamma}{2} \cos \frac{\beta + \gamma}{2} \tag{157}$$

and the evident relationship $s^2 = 1$, rewrite expression 156 as

$$\Delta = -\frac{s}{2} \left[\sin \alpha - \sin \left(s\zeta_{mnp} \right) \right] = -\frac{1}{2} \left(s \sin \alpha - \sin \zeta_{mnp} \right). \tag{158}$$

Substitute expressions 153 and 158 into 151 to obtain

$$P(\alpha) = \frac{2s_{mnp} \cos \zeta_{mnp}}{s \sin \alpha - \sin \zeta_{mnp}}. \tag{159}$$

We can show that the integral value of 150 does not depend on the sign of unit function s in expression 159. To do this, represent function 159 as the sum of its even and odd parts:

$$P(\alpha) = P^{+}(\alpha) + P^{-}(\alpha), \quad P^{\pm}(-\alpha) = \pm P^{\pm}(\alpha). \tag{160}$$

Using expressions

$$P^{\pm}(\alpha) = \frac{1}{2} \left[P(\alpha) \pm P(-\alpha) \right], \tag{161}$$

we obtain

$$P^{+}(\alpha) = \frac{s_{mnp} \sin 2\zeta_{mnp}}{\sin^2 \alpha - \sin^2 \zeta_{mnp}},$$

$$P^{-}(\alpha) = \frac{2ss_{mnp} \cos \zeta_{mnp} \sin \alpha}{\sin^2 \alpha - \sin^2 \zeta_{mnp}}, \tag{162}$$

where relationship $s^2 = 1$ is taken into account.

Now divide the contour of integration in 150 into two parts $\Gamma = \Gamma_1 \cup \Gamma_2$, where Γ_1 belongs to semiplane $\operatorname{Im} \alpha > 0$ and Γ_2 to semiplane $\operatorname{Im} \alpha < 0$. Then integral 150 can be written as

$$\psi = I_1^+ + I_2^+ + I_1^- + I_2^-, \tag{163}$$

$$I_k^\pm = \frac{1}{4\pi i} \int_{\Gamma_k} P^\pm(\alpha) \exp\left(-i\nu \sin^2 \alpha\right) d\alpha \quad \text{for} \quad k = 1, 2, \tag{164}$$

where functions $P^\pm(\alpha)$ are determined by equations 162. Replacing variable α with $-\alpha$ in integral 164 for $k = 2$, we obtain

$$I_2^+ = I_1^+, \quad I_2^- = -I_1^-. \tag{165}$$

Substitute these expressions into 163 to obtain

$$\psi = 2I_1^+. \tag{166}$$

Expressions 164 and 162 show that integral 166 does not depend on parameter s. Thus, in expression 159, we can take either of the variants $s = 1$ or $s = -1$. Take $s = 1$. Then substituting equation 159 into 150 and representing parameter ν by expression 104, we obtain

$$\psi = s_{mnp} \psi\left(\rho_{mnp}, \zeta_{mnp}\right), \tag{167}$$

where

$$\psi(\rho, \zeta) = \frac{\cos \zeta}{2\pi i} \int_\Gamma \frac{\exp\left(-\dfrac{i\pi\rho^2}{2} \sin^2 \alpha\right)}{\sin \alpha - \sin \zeta} d\alpha. \tag{168}$$

Consider some useful properties of this function. Its values at the primary and secondary shadow boundaries and at the central ray are determined by the expressions

$$\psi\left(\rho, \pm\frac{\pi}{2}\right) = 0, \quad \Psi(\rho, 0) = \frac{1}{2}, \quad \Psi(0, \zeta) = \frac{1}{2} - \frac{\zeta}{\pi}. \tag{169}$$

Its value outside the neighborhood of the central ray is determined by the asymptotic formula

$$\psi(\rho, \zeta) \sim \frac{\cot \zeta \exp\dfrac{5\pi i}{4}}{\pi\rho\sqrt{2}} + O\left(\frac{1}{\rho^2}\right) \quad \text{as} \quad \rho \to \infty, \tag{170}$$

which fails when $\zeta \to 0$. The asymptotic formula, valid as $\zeta \to 0$, is represented by the infinite series

$$\psi(\rho, \zeta) \sim \sum_{n=-\infty}^{\infty} \left[s_n^+ W\left(w_n^+\right) - s_n^- W\left(w_n^-\right)\right] \quad \text{as} \quad \rho \to \infty, \quad (171)$$

where $w_n^{\pm} = \left|\psi^{\pm} - 2\pi n\right|\rho$, $s_n^{\pm} = \text{sign}\left(\psi_n^{\pm} - 2\pi n\right)$, and $\psi^+ = \zeta$, $\psi^- = \pi - \zeta$, $0 \leq \zeta \leq \pi/2$. Imaginary and real values for argument are connected by the relationship

$$\psi\left(i|\rho|, \zeta\right) = \text{Re } \psi\left(|\rho|, \zeta\right) - i \text{ Im } \psi\left(|\rho|, \zeta\right). \quad (172)$$

Graphs of the modulus and argument of function 168 are shown in Figure 10.

By substituting expression 167 into 148, we obtain the final expression for the tip wave:

$$f_{mnp} = s_{mnp}\Phi_m H\left(\rho_{mnp}, \zeta_{mnp}\right)\exp\left(i\omega\tau_{mnp}\right), \quad (173)$$

with

$$H\left(\rho, \zeta\right) = W\left(\rho\right)\psi\left(\rho, \zeta\right), \quad (174)$$

$$\rho_{mnp} = \sqrt{\frac{2\omega\left(\tau_{mnp} - \tau_m\right)}{\pi}}, \quad \zeta_{mnp} = \arcsin\sqrt{\frac{\tau_{mnp} - \tau_{mn}}{\tau_{mnp} - \tau_m}}, \quad (175)$$

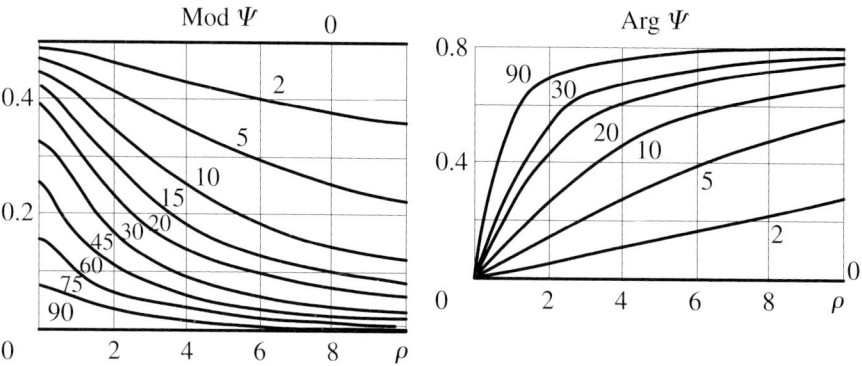

Figure 10. Function $\Psi(\rho, \zeta)$. Curve labels are values of ζ (Klem-Musatov, 1981).

and

$$s_{mnp} = +1 \quad \text{for} \quad M \in \Omega_{mnp}^+, \quad \text{and} \quad s_{mnp} = -1 \quad \text{for} \quad M \in \Omega_{mnp}^-,$$

$$(176)$$

where positions of regions $\Omega_{mnp}^{\pm}$ are shown in Figure 5.

Diffraction at a vertex

It follows from the last expression in 169 that any point of the central ray $\rho_{mnp} = 0$ is an essential singular point of the tip wavefield because its value depends on direction ζ_{mnp}, along which this point is approached. We will show here that the total wavefield at points of the central ray does not depend on the direction of approach, i.e., it is determined uniquely. To do this, consider the wavefield caused by diffraction at the vertex.

An interface in a small neighborhood of the vertex forms a curvilinear polyhedron (Figure 11). A single wave f_m is formed by the reflection/transmission at an individual face of the polyhedron. Specify this face by index m of the corresponding wave f_m.

Denote the domains of waves f_m and f_{mn} by Ω_m and Ω_{mn}, respectively, and introduce the unit functions

$$\delta_m = \begin{cases} 1 & \text{for} \quad M \in \Omega_m, \\ 0 & \text{for} \quad M \notin \Omega_m, \end{cases} \qquad \delta_{mn} = \begin{cases} 1 & \text{for} \quad M \in \Omega_{mn}, \\ 0 & \text{for} \quad M \notin \Omega_{mn}, \end{cases} \qquad (177)$$

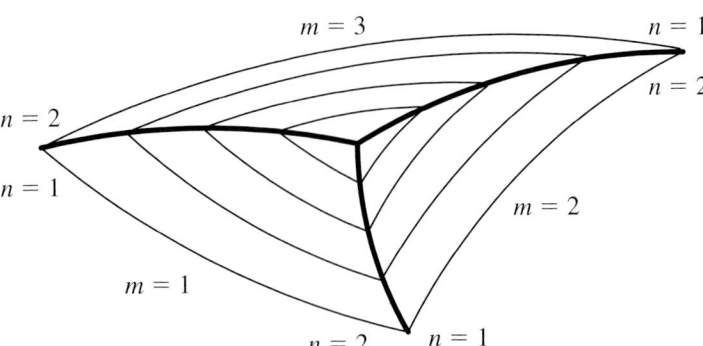

Figure 11. Interface in a small neighborhood of the vertex. The index m denotes a face, and the index n denotes an edge of the mth face (Klem-Musatov, 1994).

where M is the current point of space. There are obvious relationships between unit functions 58 of Chapter 3 and 176 above:

$$s_{mnp} = -s_{mn} \quad \text{for} \quad M \in \Omega_{mn},$$
$$s_{mnp} = +s_{mn} \quad \text{for} \quad M \notin \Omega_{mn} \tag{178}$$

or, equivalently,

$$s_{mnp} = -s_{mn}, \quad \text{if} \quad \delta_{mn} = 1, \quad \text{and} \quad s_{mnp} = +s_{mn}, \quad \text{if} \quad \delta_{mn} = 0. \tag{179}$$

Then wavefield 1 resulting from the polyhedral interface can be written as

$$f = \sum_m \left[\delta_m f_m + \sum_n \left(\delta_{mn} f_{mn} + \sum_p f_{mnp} \right) \right]. \tag{180}$$

We will limit ourselves by considering only the contribution from an individual face, omitting the symbol of summation over index m for brevity. Each wave f_{mn} is caused by one of two edges ($n = 1$ or $n = 2$) of the mth face, and each wave f_{mn} produces only one tip wave f_{mnp}. Then instead of expression 180, we have

$$f = \delta_m f_m + \sum_{n=1}^{2} \left(\delta_{mn} f_{mn} + f_{mnp} \right). \tag{181}$$

Consider the case when point M lies on the central ray. From equation 57 of Chapter 3 with $\tau_{mn} = \tau_m$, it follows that

$$f_{mn} = \frac{s_{mn} f_m}{2}, \tag{182}$$

and from equations 173 and 169 with $\rho_{mnp} = 0$, it follows that

$$f_{mnp} = \frac{s_{mnp} f_m \psi(0, \zeta_{mnp})}{2} = s_{mnp} f_m \left(\frac{1}{4} - \frac{\zeta_{mnp}}{2\pi} \right). \tag{183}$$

Substituting expressions 182 and 183 into 181, we obtain the value of the total wavefield at the central ray,

$$f = f_m \left\{ \delta_m + \sum_{n=1}^{2} \left[\frac{\delta_{mn} s_{mn}}{2} + s_{mnp} \left(\frac{1}{4} - \frac{\zeta_{mnp}}{2\pi} \right) \right] \right\}$$
$$= f_m \left[\delta_m + \sum_{n=1}^{2} \left(x - \frac{s_{mnp} \zeta_{mnp}}{2\pi} \right) \right], \tag{184}$$

where

$$x = \frac{\delta_{mn} s_{mn}}{2} + \frac{s_{mnp}}{4}. \tag{185}$$

Express s_{mnp} in terms of s_{mn}, with $\delta_{mn} = 0$ and $\delta_{mn} = 1$ according to equations 179, so that

$$x = \frac{s_{mn}}{4}. \tag{186}$$

Substitute this expression into 184 to obtain

$$f = f_m \left[\delta_m + \sum_{n=1}^{2} \left(\frac{s_{mn}}{4} - \frac{s_{mnp} \zeta_{mnp}}{2\pi} \right) \right] \tag{187}$$

or, equivalently,

$$f = \frac{f_m \varepsilon_m}{2\pi}, \tag{188}$$

$$\varepsilon_m = 2\pi \delta_m + \gamma_m, \tag{189}$$

$$\gamma_m = \xi_{m1} + \xi_{m2}, \tag{190}$$

and

$$\xi_{mn} = \frac{s_{mn} \pi}{2} - s_{mnp} \zeta_{mnp}. \tag{191}$$

Expression 188 describes the value of the wavefield at the central ray. However, it has a coefficient, 189, expressed in terms of the coordinate ζ_{mnp}. Formally, this means quantity 188 depends on the direction of approach to the central ray. In fact, we will show that it does not. To do this, we rewrite expression 189 in a form convenient for geometric interpretation. We will illustrate geometric relationships by illustrating the positions of different lines of intersections of the coordinate surfaces $\psi^{\pm} = $ constant with the reflected/transmitted wavefront. Because we are considering the case of $\sigma \to 0$, it is possible to approximate all surfaces by their tangent planes at $\sigma = 0$. Then all lines of intersection of coordinate surfaces can be approximated by straight lines.

Write equation 191 as

$$\xi_{mn} = s_{mn}\beta_{mn}, \quad \beta_{mn} = \pi/2 - s_{mn}s_{mnp}\zeta_{mnp}. \tag{192}$$

Because

$$s_{mn}s_{mnp} = \begin{cases} -1 & \text{when} \quad M \in \Omega_{mn}, \\ +1 & \text{when} \quad M \notin \Omega_{mn}, \end{cases} \tag{193}$$

the following inequalities hold true:

$$0 \le \beta_{mn} \le \pi, \quad |\xi_{mn}| < \pi. \tag{194}$$

One can see that quantity β_{mn} is the dihedral angle between surface $\psi^{\pm} = \pi/2$ and surface $\psi^{\pm} =$ constant containing point M (Figure 12).

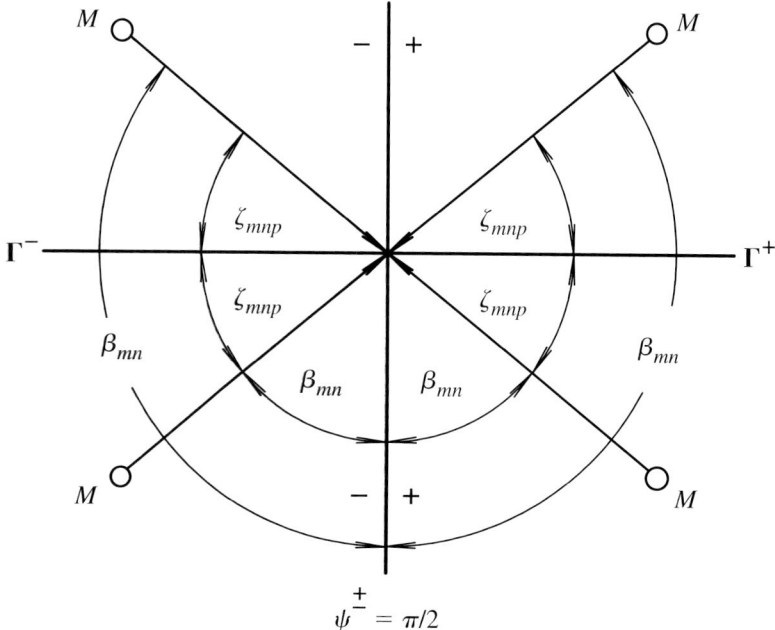

Figure 12. Geometric interpretation of the quantity β_{mn} with different directions of approach to the central ray. This figure depicts the tip wavefront in a small vicinity of the central ray. M is the observation point. Signs $+$ and $-$ indicate values of the unit function s_{mn}.

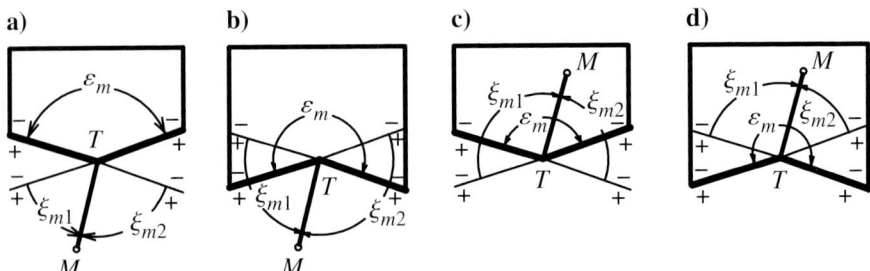

Figure 13. Geometric interpretation of equation 189, depicting the domain Ω_m of wave f_m in a small vicinity of the central ray, with (a, b) $M \in \Omega_m$, and (c, d) $M \notin \Omega_m$. ε_m is the dihedral angle between the primary shadow boundaries of wave f_m. Signs + and − indicate values of unit function s_{mn}.

If $M \notin \Omega_m$, quantity 189 has the form

$$\varepsilon_m = \gamma_m = s_{m1}\beta_{m1} + s_{m2}\beta_{m2}. \tag{195}$$

Figure 13 illustrates that this quantity is always positive and corresponds to the angular size of the primary illuminated zone Ω_m of wave f_m.
  If $M \notin \Omega_m$, we have

$$\varepsilon_m = 2\pi + \gamma_m = 2\pi - \left(\beta_{m1} + \beta_{m2}\right). \tag{196}$$

Figure 13 shows that γ_m is always negative. Quantity $-\gamma_m$ corresponds to the complement of the angular size of the primary illuminated zone Ω_m to 2π.
  Thus, ε_m corresponds to the dihedral angle between the primary shadow boundaries of wave f_m. This angle must be taken within the primary illuminated zone Ω_m. One can see that the value of the wavefield at the central ray, including edge and tip diffraction 188, does not depend on the direction of approach to the central ray. Remember that the value of the wavefield, taking into account edge diffraction in the case of the smooth edge, is given by relationship 137 of Chapter 2. Expression 188 above can be regarded as the generalization of that relationship to the case of a nonsmooth edge. In the particular case where $\varepsilon_m = \pi$ (i.e., in the case of a smooth edge), expression 188 takes the form 137 of Chapter 2.

Vector waves

Let the reflected/transmitted wave f_m be a vector quantity 59 of Chapter 3. Then the vector edge wave can be written in the form 62 and 63 of Chapter 3.

To find the vector tip wave, decompose the latter over the coordinate basis of the reflected/transmitted wave 60 of Chapter 3,

$$f_{mnp} = \sum_{q=1}^{3} \mathbf{j}_q f^q_{mnp}, \quad f^q_{mnp} = \psi^q_{mnp} \exp\left(i\omega\tau_{mnp}\right), \qquad (197)$$

where ψ^q_{mnp} are unknown scalars. To find these scalars, use the same approach as in the scalar case, i.e., expression 173 above. This allows us to find three scalar functions

$$f^q_{mnp} = s_{mnp} \psi^q_m H\left(\rho_{mnp}, \zeta_{mnp}\right)\exp\left(i\omega\tau_{mnp}\right) \quad \text{for} \quad q = 1, 2, 3. \tag{198}$$

Substituting these expressions into 197, we again obtain equation 173, where Φ_m is the vector amplitude of wave 59 of Chapter 3. Druzhinin (1991) shows that these equations are also valid in anisotropic media.

This result can be interpreted in the same way as for the edge wave in Chapter 3, in the section titled "Vector waves." Let $\mathbf{p}_{mnp}$ be the unit polarization vector of tip wave f_{mnp}. In accordance with the general theory, this vector must coincide with the unit vector $\mathbf{e}_{mnp}$ of the tangent to the tip wave (for a longitudinal wave) or must be perpendicular to $\mathbf{e}_{mnp}$ (for a transverse wave). However, vector $\mathbf{p}_{mnp}$ coincides with polarization vector $\mathbf{p}_m$ of the reflected/transmitted wave. In fact, within the boundary layer, the difference between $\mathbf{p}_{mnp}$ and $\mathbf{p}_m$ is negligibly small. The two vectors thus are considered to be equal.

Chapter 5

Examples and Applications

Edge- and tip-wave theories were developed during a time when computational power was not readily available for verification by comparing with full wave solutions. However, physical modeling of wave propagation was common in several Soviet laboratories, including the Institute of Geophysics in Novosibirsk, where the initial theory and algorithms were developed (Klem-Musatov et al., 1972, 1975, 1976, 1982; Aizenberg and Klem-Musatov, 1980; Aizenberg, 1982). The first section of this chapter reviews experiments made by Russian scientists to compare their theoretical calculations against experimental data in simple 2D and 3D models (Klem-Musatov, 1980; Landa and Maksimov, 1980; Luneva and Kharlamov, 1990).

Because theory and applications of edge and tip waves were published in Western journals (Klem-Musatov and Aizenberg, 1984, 1985, 1989), several groups pursued their own implementation, e.g., Pajchel et al. (1987, 1988, 1989) in Norway, Hoffmann et al. (1993) and Klaeschen et al. (1994) in Germany, Hron and Chan (1995) in Canada, and Wang and Waltham (1995) in the United Kingdom. As ray-method applications developed as tools in geophysical prospecting, edge-wave theory was discovered to be a convenient remedy for limitations of the ray approach in handling model discontinuities. We devote the second section of this chapter to one of the first practical implementations of edge-wave theory: the 2D software package of Pajchel et al. (1987). This implementation was used widely for practical exploration problems in the North Sea, where discontinuities in geologic structures and diffractions are common features of seismic sections.

Edge-wave theory fails where the ray-theory field changes rapidly, e.g., in the vicinity of caustic zones. These shortcomings are eliminated in the tip-wave superposition method (TWS), which provides accurate seismograms for models with curved boundaries and faults. In the third section of this chapter, we review the algorithm for the TWS and its prototype software implementation for a model consisting of two layers separated by an interface of arbitrary complexity. Real surfaces from the North Sea containing faults and synclines are used for numerical experiments of seismic surveys.

Aizenberg et al. (1992a, 1992b) highlight some attractive features of the TWS approach in 3D seismic modeling. A recent paper by Klem-Musatov et al. (2004) provides a rigorous theoretical basis for improvement of the TWS approach. Ayzenberg et al. (2007a) offer a review of the theory and its practical implementation for a single interface. Klem-Musatov et al. (2005a) provide the theoretical background for its extension to multilayer media. Ayzenberg et al. (2007b) briefly outline its practical implementation.

Synthetic seismograms for simple subsurfaces and comparison with physical experiments

Here we will demonstrate the physical meaning of diffraction phenomena in the boundary-layer approximation for simple models of interfaces. Such models allow us to demonstrate separately the ray method and diffraction components of the wavefield and to see the effects of their superposition. They also let us compare the boundary-layer approximation with the more accurate description and with experimental data.

Diffraction on a half plane

In this case, an individual edge wave can be observed in the boundary layer. Figure 1 shows the observation system (Klem-Musatov and Aizenberg, 1985). The source (marked with an asterisk) excites a nonstationary spherical wave. The line of receivers (illustrated with circles) records the wave reflected from the half plane. Because the reflecting surface is bounded by the edge, the reflected wave has a shadow zone for $x > 1.15$ km (Figure 2a). Diffusion of energy from the shadow boundary appears in the form of the diffracted edge wave propagating from the edge (Figure 2b). The total wavefield is formed by superposition of the reflected and edge-diffracted waves (Figure 2c).

Figure 3 shows framed details of the seismograms from Figure 2. The edge wave's phase inversion occurs at the shadow boundary (Figure 3b). Its amplitude rapidly decays with distance from the shadow boundary (Figure 3b). The superposition of two discontinuous waves (Figure 3a and b) is continuous at the shadow boundary (Figure 3c). The intensity of the diffraction effect is comparable with the reflection in the boundary layer, 0.85 km $\leq x \leq 1.45$ km. Outside the boundary layer, the diffraction effect is practically invisible (compare the traces $x = 0.8$ km in Figure 3a and 3c).

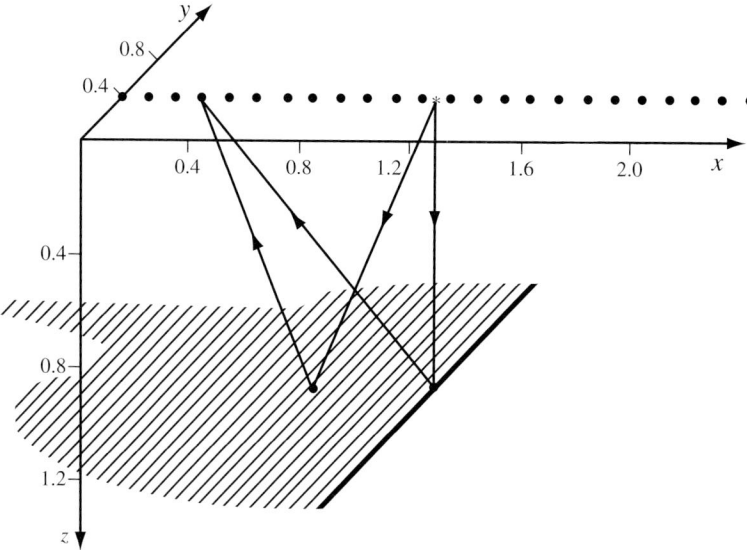

Figure 1. Model of a half plane. The reflecting interface coincides with the shaded region (from Figure 2, Klem-Musatov and Aizenberg [1985]). © 1985, *Journal of Geophysics.* Used with kind permission of Springer Science and Business Media.

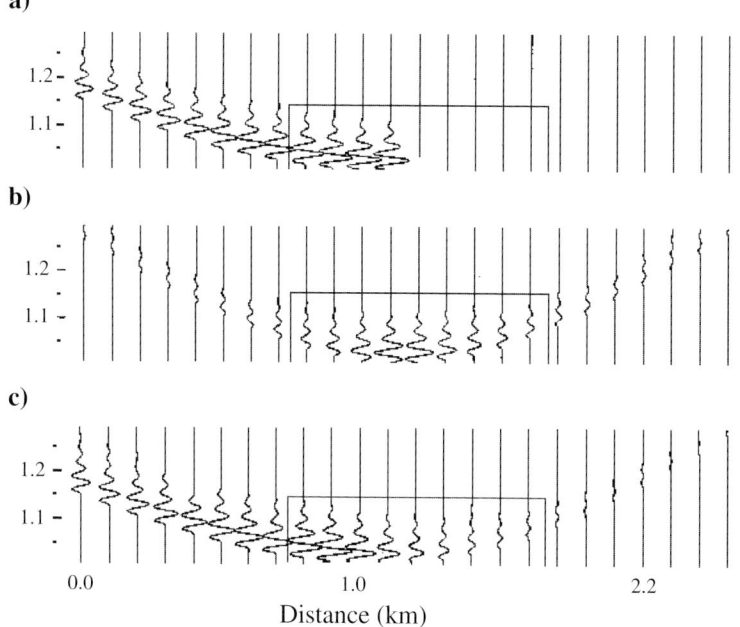

Figure 2. Theoretical seismograms for model of the half plane shown in Figure 1: (a) reflected wave, (b) edge wave (twice enlarged), and (c) total field (from Figure 3, Klem-Musatov and Aizenberg [1985]). © 1985, *Journal of Geophysics.* Used with kind permission of Springer Science and Business Media.

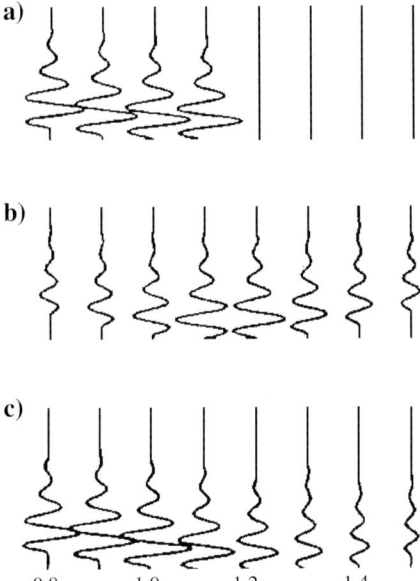

Figure 3. Detail panels of the theoretical seismograms from Figure 2: (a) reflected wave, (b) edge waves (twice enlarged), and (c) total field (detail from Figure 3, Klem-Musatov and Aizenberg [1985]). © 1985, *Journal of Geophysics.* Used with kind permission of Springer Science and Business Media.

Figure 4 compares theoretical edge-wave amplitude graphs with experimental data obtained by physical modeling (Klem-Musatov et al., 1972). The reflecting half plane is placed in a liquid. Profiles of observation (on the liquid's surface) form different angles α with projection of the edge on the observation surface (the graphs depict values of α). Receivers are placed in the shadow zone to detect only the edge wave. One can see that the boundary-layer approximation describes the edge-diffracted wave with sufficient accuracy.

Diffraction on a wedge

We consider ray tracing in a simple 2D subsurface model consisting of three layers, each

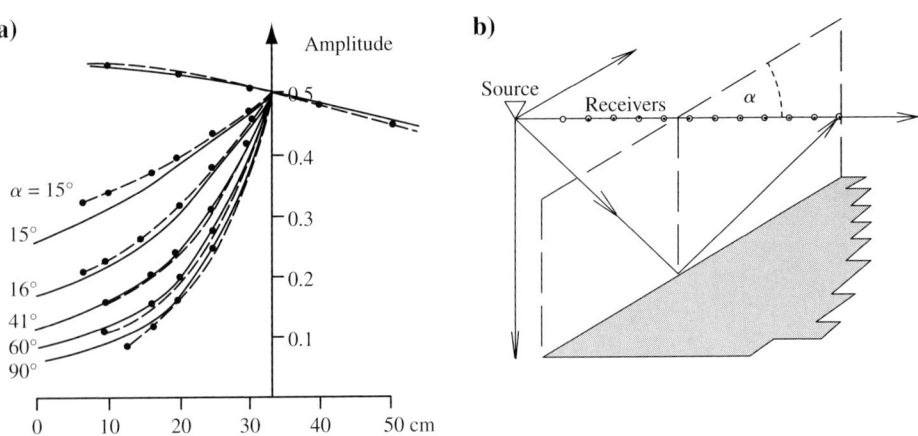

Figure 4. (a) Experimental (left, dashed lines) and theoretical (solid lines) amplitude graphs of the edge wave from a rectilinear edge. The top graphs correspond to the reflection from the infinite plane. (b) The experimental setup (from Klem-Musatov et al., 1972).

with homogeneous material properties, separated by a flat seabed and a deeper broken interface forming a wedge, as illustrated in Figure 5 (left column). Source and receivers are located at the sea surface. Reflecting rays are traced from the deep reflector and back to the surface by using standard ray-tracing techniques (Červený et al., 1977). No reflecting rays can be traced in the shadow zone, and a corresponding gap appears in the seismogram (R1 and R2). Because discontinuity in the response is not realistic, the objective here is to remedy the limitation of the ray theory by filling the gap, using edge waves.

The wedge is made by the intersection of two line segments, each forming a diffracting edge with response D1 and D2 (Figure 5, right column). Because this diffraction point arises from the intersection of two reflector segments, two edge waves will contribute to the diffraction response. We combine the reflection response computed from conventional ray tracing (R1 + R2) and the diffraction response from edge-wave theory (D1 + D2), both being discontinuous. Thus we obtain the complete response, which is smooth (see Figure 5, middle column). The results dem-

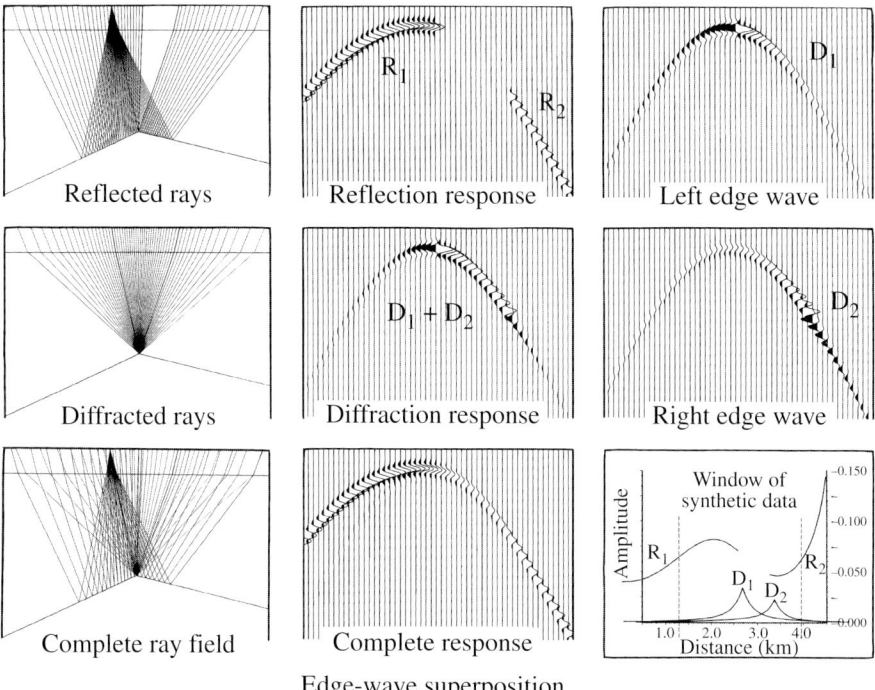

Edge-wave superposition

Figure 5. Basic concepts of the edge-wave superposition method demonstrated in a model with a broken reflector buried below the seabed (left panel). Each contribution to the wavefield, R1, R2, D1, and D2 (all of which are discontinuous functions), are superimposed to form a complete response, which is smooth (middle panel) (from Pajchel et al., 1988).

onstrate clearly the advantage of the edge-wave approach, which preserves the capability of ray-based modeling to separate the wavefield into individual arrivals.

This approach is so simple because only the edge-wave components that smooth discontinuities of reflected waves at the shadow boundaries are considered. Those components describe the edge wave only in the vicinities of shadow boundaries — accuracy depends on the distance from the edge. The more accurate, uniform description of the edge wave, which is valid simultaneously in boundary layers and deep shadow zones, demands significantly more computational effort (Klem-Musatov, 1980, 1994). The following comparison (Figure 6) with the uniform description allows us to estimate the range of applicability of the boundary-layer approximation.

The system of polar coordinates (r, θ) with the origin at the corner point is oriented so that line $\theta = 0$ coincides with the axis of symmetry of the wedge in the upper layer. The angle of the wedge in the upper layer is $270°$. The incident-plane acoustic wave propagates in the upper media in the direction $\theta = 180°$. Upper media are characterized by wave velocity c_1 and density ρ_1. Corresponding properties of the lower layer are c_2 and ρ_2.

Figure 6 depicts distribution of the edge-wave amplitude modulus along the edge wavefront for the different medium properties and distances from the corner point. Distance is characterized by quantity $k_1 r = 2\pi r/\lambda_1$, $k_1 = \omega/c_1$, where λ_1 is the wavelength and ω is angular frequency. All curves have their maximum at the shadow boundary where edge-wave amplitude equals one-half of reflected-wave amplitude. Comparison shows that the relative error of the boundary-layer approximation diminishes with distance from the edge and is sufficiently small for seismic-prospecting situations (curves C in Figure 6).

Diffraction on a pinched-out layer

A subsurface model consists of three homogeneous regions (Figure 7) with the elastic parameters

$$\rho_1 = 2 \text{ g/cm}^3, \ v_{p1} = 2 \text{ km/s}, \ v_{s1} = 1.25 \text{ km/s},$$

$$\rho_2 = 2.4 \text{ g/cm}^3, \ v_{p2} = 2.5 \text{ km/s}, \ v_{s2} = 1.5 \text{ km/s},$$

$$\text{and } \rho_3 = 1.8 \text{ g/cm}^3, \ v_{p3} = 1.75 \text{ km/s}, \ v_{s3} = 1.05 \text{ km/s},$$

in which v_{ij} is the velocity of the longitudinal ($i = p$) or transverse ($i = s$) wave, and ρ_j is the media density when j is the medium index. The source

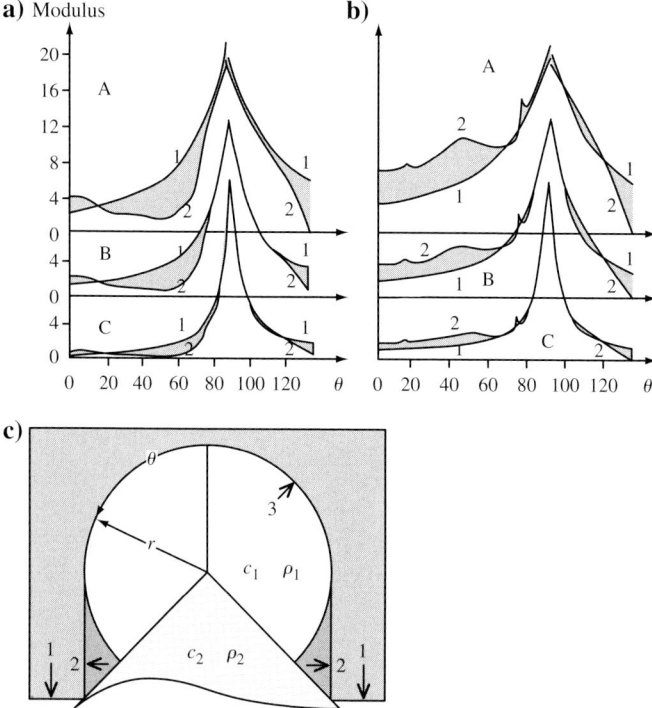

Figure 6. Comparison of the boundary-layer approximation (curves 1) with the uniform asymptotic description (curves 2) of the edge-wave amplitude modulus. (a) $c_1/c_2 = 1/2$, $\rho_1/\rho_2 = 1/2$;. (b) $c_1/c_2 = 2/1$, $\rho_1/\rho_2 = 2/1$; (A) $k_1 r = 12.5 \left(r \approx 2\lambda_1 \right)$; (B) $k_1 r = 50 \left(r \approx 8\lambda_1 \right)$; (C) $k_1 r = 250 \left(r \approx 40\lambda_1 \right)$. $k_1 = \omega/c_1$, λ_1 represents the wavelength. (c) Wavefronts at the wedge: 1 is incident wave; 2, reflected waves; 3, edge wave (from Klem-Musatov, 1980).

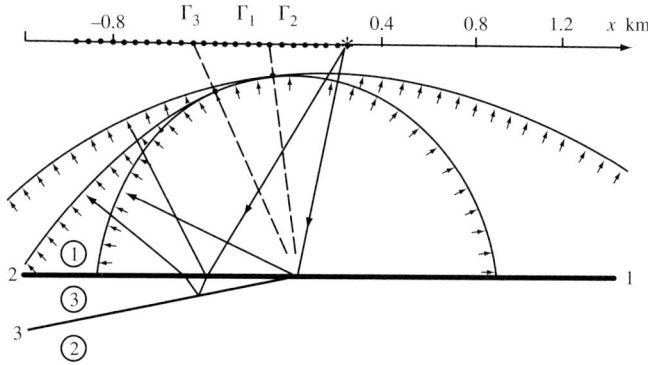

Figure 7. Model of pinch-out and wavefronts in cross section with projection of the observation system. The profile of observation is perpendicular to the edge. The angle of the pinch-out is 5° (from Figure 7, Klem-Musatov and Aizenberg [1985]). © 1985, *Journal of Geophysics*. Used with kind permission of Springer Science and Business Media.

(indicated in Figure 7 by the asterisk) and receivers (circles) are located at the surface. The profile of observation is perpendicular to the edge. Here, theoretical seismograms show the vertical component of longitudinal waves only.

There are multiple reflections inside the pinched layer. As a result, many reflected waves have sharp shadow boundaries at the profile of observation. Only three of them appear in Figure 7. Their shadow boundaries are indicated with dashed lines (Γ_1, Γ_2, relating to the primary reflections from interfaces 1 and 2, and Γ_3 relating to the reflection from interface 3 with transmission through interface 2).

Figure 8 depicts the wavefields (Klem-Musatov and Aizenberg, 1985). Detail panels of the seismograms show the mechanism of smoothing

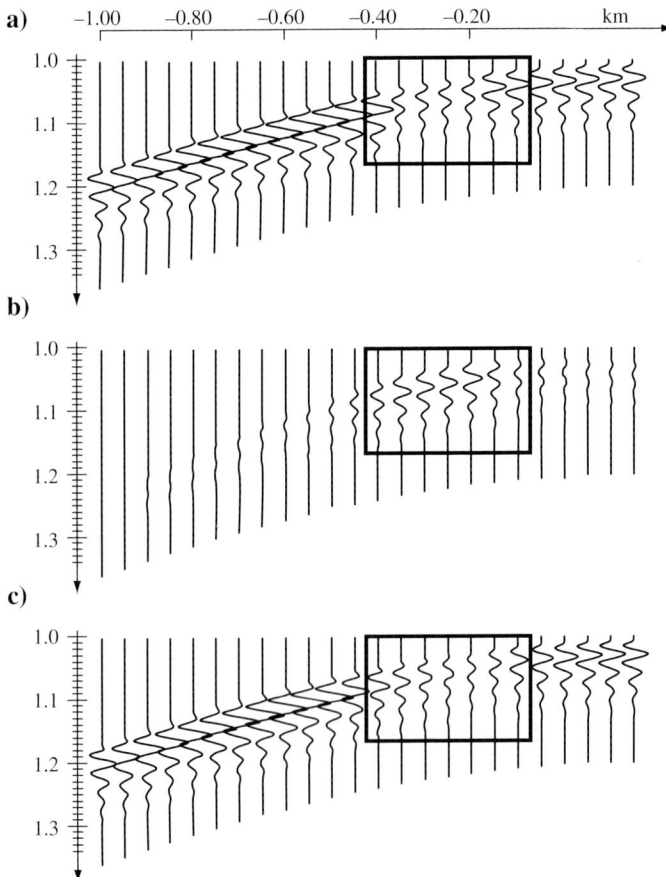

Figure 8. Theoretical seismograms for pinch-out model: (a) reflected waves, (b) edge waves, and (c) total field. Details in the indicated windows are depicted in Figure 9 (from Figure 8, Klem-Musatov and Aizenberg [1985]). © 1985, *Journal of Geophysics.* Used with kind permission of Springer Science and Business Media.

a)

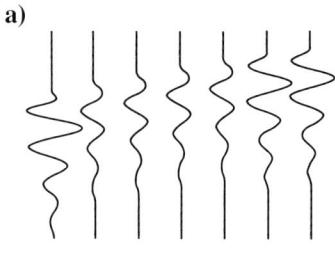

b)

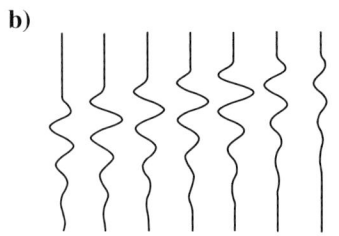

c)

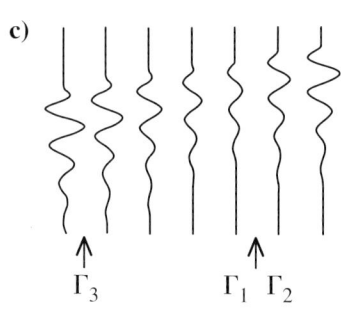

Γ_3 Γ_1 Γ_2

Figure 9. Details from seismograms in Figure 8, indicating the compensation of the discontinuities of reflected waves on the shadow boundaries (shown by arrows) resulting from the phase inversion of edge waves in theoretical seismograms: (a) reflected waves, (b) edge waves, and (c) total field (detail from Figure 8, Klem-Musatov and Aizenberg [1985]). © 1985, *Journal of Geophysics.* Used with kind permission of Springer Science and Business Media.

reflected-wave discontinuities (Figure 9a) by phase inversion of the edge waves (Figure 9b) on shadow boundaries.

The jump of the reflected wavefield in Figure 9a at shadow boundaries Γ_1 and Γ_2 occurs because of the abrupt change of the reflection coefficient along the horizontal interface at the edge. Because reflections from interfaces 2 and 3 overlap, shadow boundary Γ_3 appears in Figure 9a as a jump of the reflected wavefield on the smooth background. As the neighborhoods of the shadow boundaries overlap, all three edge waves contribute significantly to the common diffracted wave (Figure 9b).

Figure 10 shows experimental and theoretical amplitude graphs for different angles of the pinched layer, obtained by 2D ultrasound modeling (Klem-Musatov, 1980). Elastic parameters of the model are $v_1 = v_2 = 2.2$ km/s (organic glass) and $v_3 = 1.7$ km/s (paraffin). The source (indicated by the black triangle in Figure 10) and receivers are located at the surface. Notice that the shift of amplitude graphs below the mark $A = 0.5$ at $x = 0$ in Figure 10c results from overlapping of the boundary layers in waves reflected from top and bottom interfaces of the pinched-out layer.

Diffraction on a small-throw fault

A pair of half planes parallel to the observation surface is the simplest model of a small-throw fault (Figure 11b). The seismic trace is put in cor-

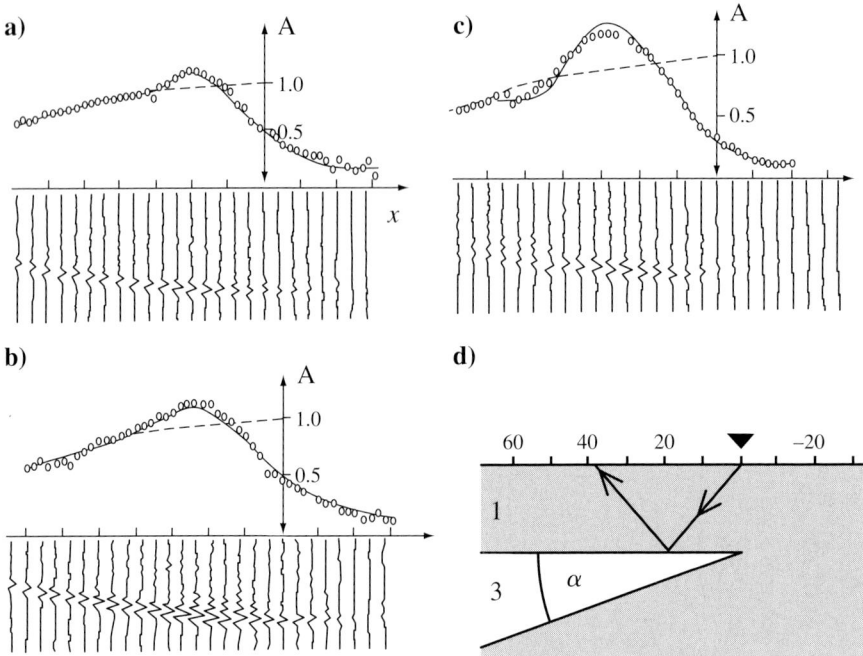

Figure 10. Normalized experimental seismograms, experimental amplitude graphs (circles), and theoretical amplitude graphs (solid lines) in the pinch-out model for (a) $\alpha = 90°$, (b), $\alpha = 20°$, and (c), and $\alpha = 5°$, where α is the angle of the pinch-out. The amplitude graph of the wave reflected from the horizontal part of the interface is indicated by a dashed line. (d) The source of oscillations (its position is shown with the black triangle) is above the edge (from Klem-Musatov et al., 1975).

respondence to the common depth point. In Figure 11a, Klem-Musatov et al. (1976) illustrate theoretical seismograms including longitudinal reflected waves and diffractions for the profile oriented across the strike of the fault. In this example, Δh is the vertical fault displacement and λ is the predominant wavelength. The presence of the fault is reflected in the form of a pronounced region of anomalous decrease of the intensity of oscillations.

Figure 12 shows experimental and theoretical seismograms of longitudinal reflected waves for the 2D model, with $\Delta h = \lambda/4$ from Landa and Maksimov (1980). The observation system is of the common-depth-point flank type with a sixfold overlap. Each seismogram corresponds to the fixed position of the source.

a)

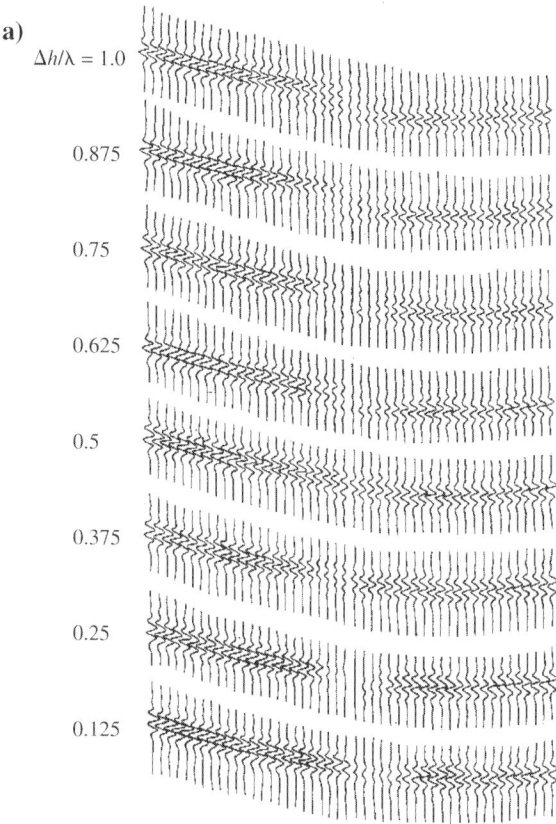

Figure 11. (a) Theoretical seismograms of reflections from (b) an interface disrupted by a small-throw fault. Δh is the vertical displacement, and λ is the predominant wavelength (from Klem-Musatov et al., 1976).

b)

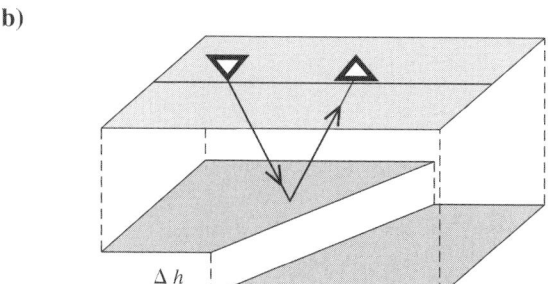

Diffraction on a sector

Figure 13 illustrates location of the interface (Klem-Musatov and Aizenberg, 1985). The diffracting edge is formed by two semi-infinite rectilinear lines with the common point (vertex). The source (marked by

a) Experiment   **b)** Theory

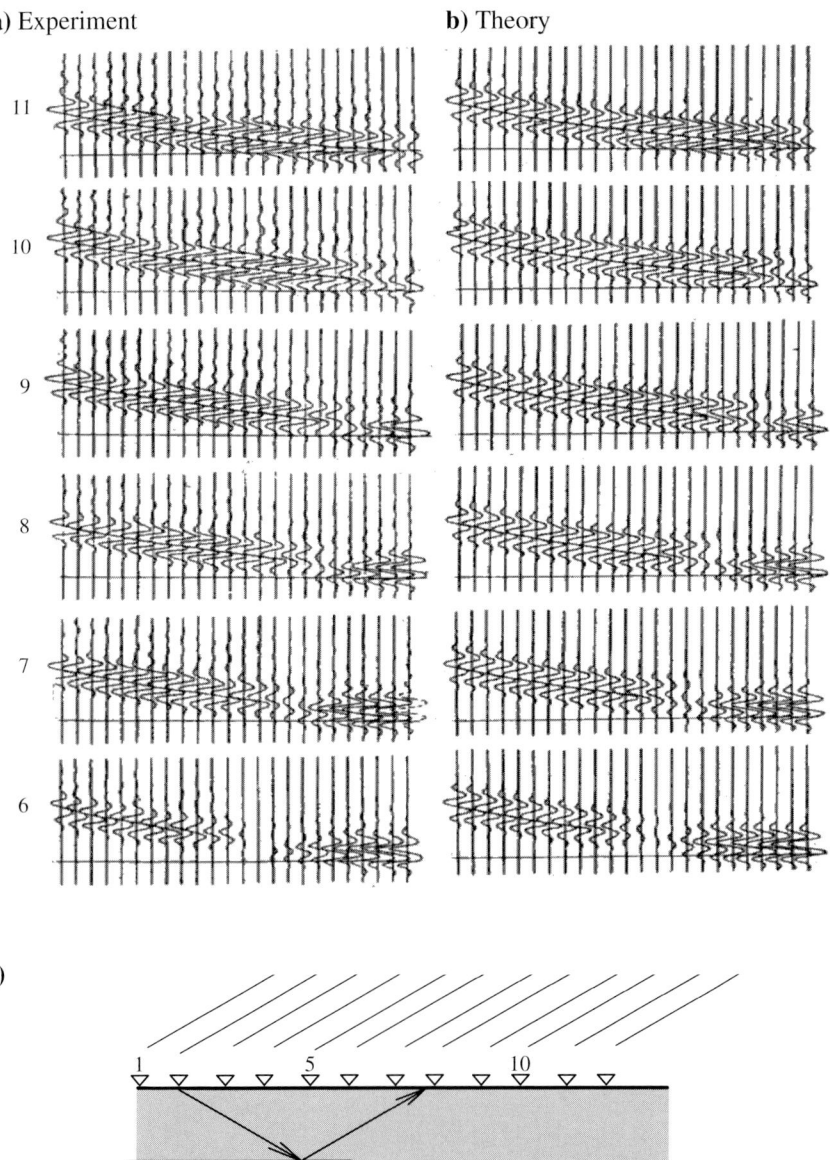

Figure 12. (a) Experimental seismogram and (b) theoretical seismogram for the model of a small-throw fault $\Delta h = \lambda/4$ with (c) the observation plan. Shot numbers are indicated on the left side (from Landa and Maksimov, 1980).

the asterisk) and two profiles of receivers (indicated by lines) are located at the surface. The total wavefield is represented as a superposition of the reflected wave f_1, two edge waves f_{1n} ($n = 1, 2$), and two tip waves f_{1n1} ($n = 1, 2$). Primary shadow boundaries are marked by indices 11 and 12,

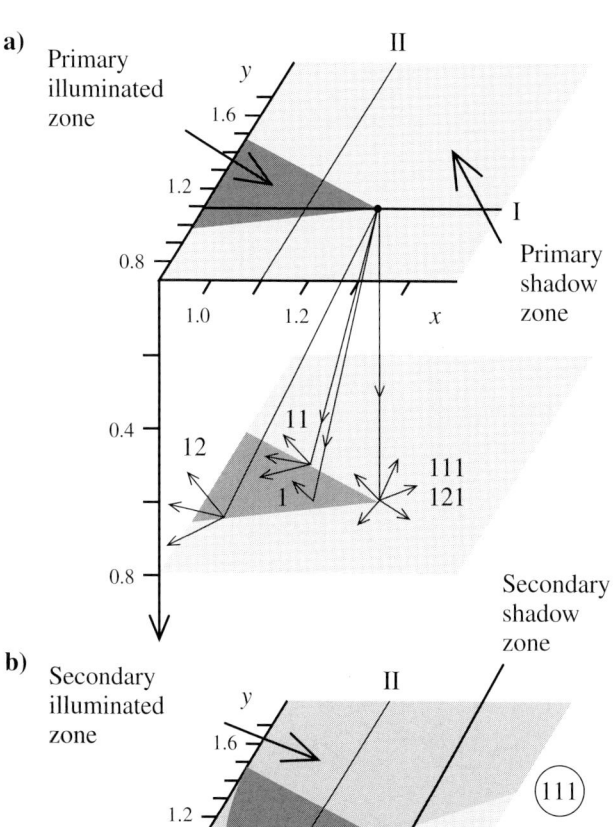

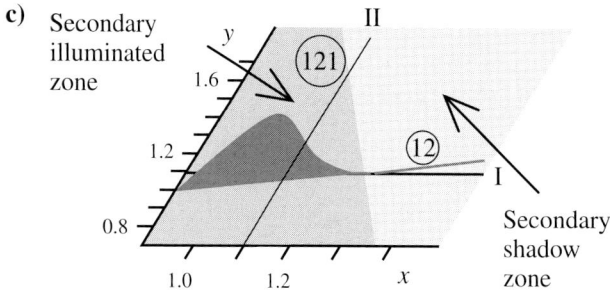

Figure 13. Scheme of reflection from the sector with positions of illuminated and shadow zones in (a) reflected and (b, c) diffracted waves. I and II are the observation profiles (modified from Figure 11, Klem-Musatov and Aizenberg [1985]). © 1985, *Journal of Geophysics*. Used with kind permission of Springer Science and Business Media.

and the secondary shadow boundaries are marked by indices 111 and 121. Figures 14a and 15a depict reflected wave for two profiles, denoted as I and II, respectively. Figures 14b and 15b illustrate the edge wavefields scattered from both edges. Figures 14c and 15c illustrate the tip waves. Figures 14d and 15d depict the total wavefields. Detail panels of the seismograms are enlarged in Figures 16 and 17.

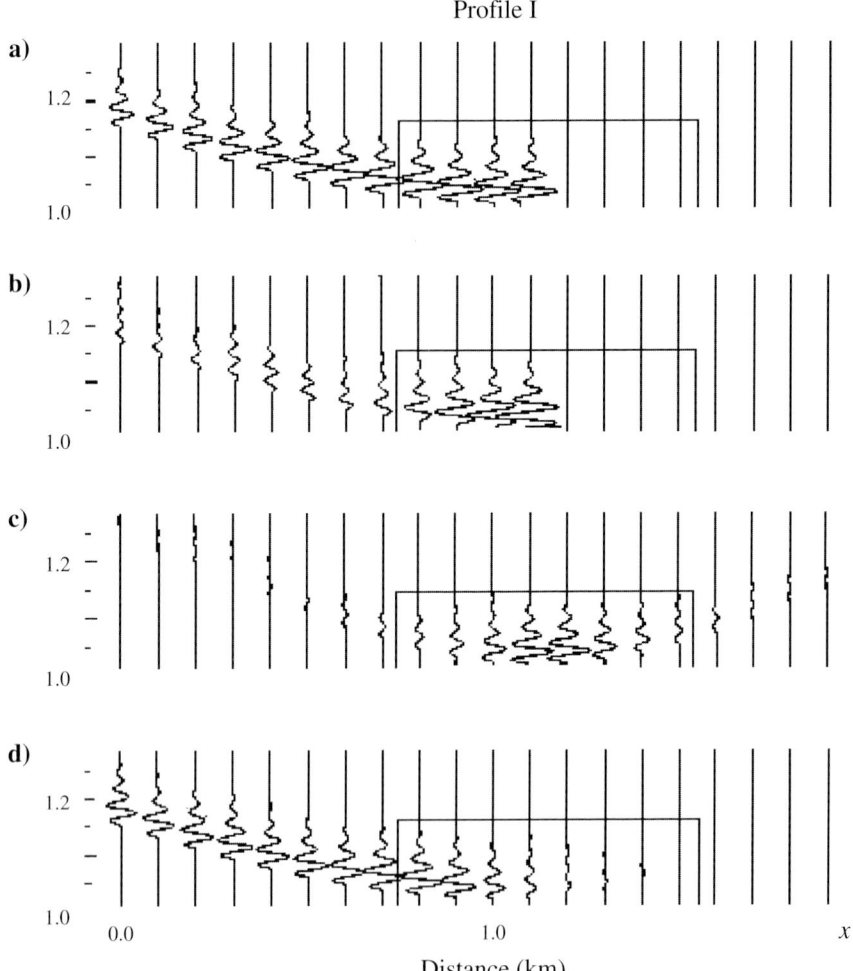

Figure 14. Theoretical seismograms for sector model (profile I): (a) reflected wave, (b) edge waves (1.5 times enlarged), (c) tip waves (7.5 times enlarged), and (d) total field. Details in the indicated windows are illustrated in Figure 16 (from Figure 12, Klem-Musatov and Aizenberg [1985]). © 1985, *Journal of Geophysics*. Used with kind permission of Springer Science and Business Media.

Relative positions of the source and observation profile I are chosen so that primary and secondary shadow zones coincide for $x > 1.1$. Thus, the seismogram for $x > 1.1$ allows us to observe the vertex wave. Smoothing of discontinuities at the primary and secondary shadow boundaries can be seen in profile II. Figures 15a and 17a illustrate the discontinuity of the reflected wave.

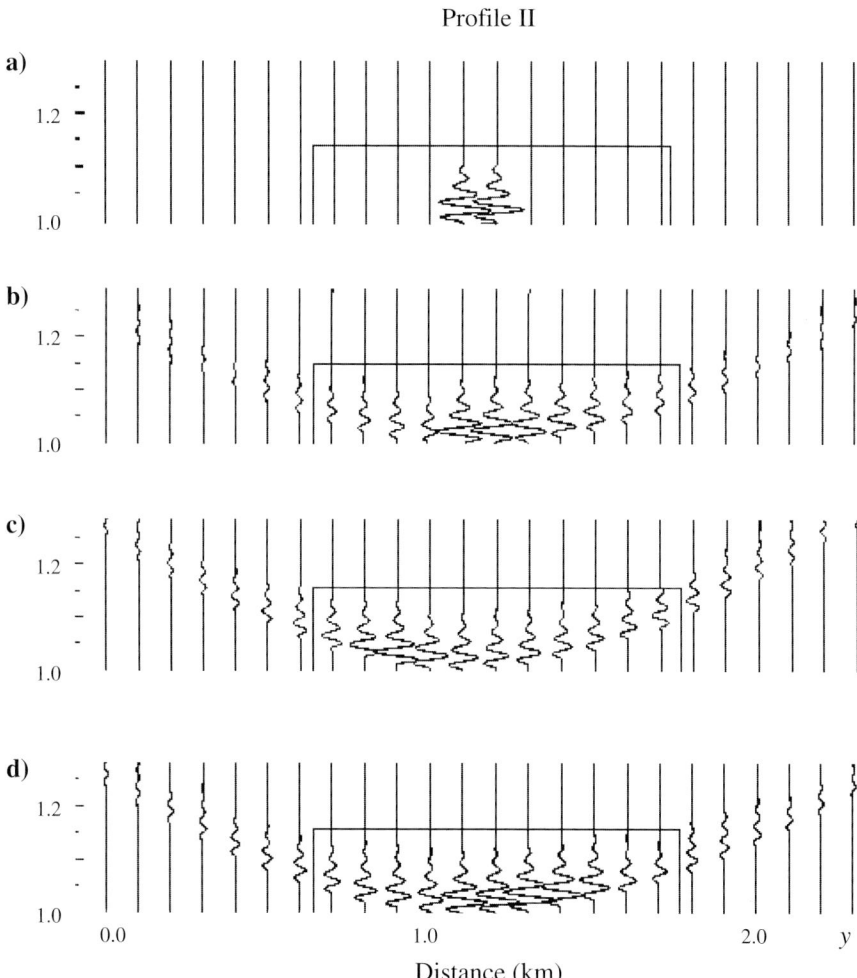

Figure 15. Theoretical seismograms for sector model (profile II): (a) reflected wave (twice diminished), (b) edge waves (1.66 times enlarged), (c) tip waves (enlarged four times), and (d) total field. Details in the indicated windows are depicted in Figure 17 (from Figure 13, Klem-Musatov and Aizenberg [1985]). © 1985, *Journal of Geophysics*. Used with kind permission of Springer Science and Business Media.

Figure 17b illustrates the phase inversion of edge waves at the primary shadow boundaries 12 ($y = 1.06$ km) and 11 ($y = 1.24$ km). Figures 15c and 17c show tip waves. One can see the phase inversion of the tip waves at secondary shadow boundaries 111 ($y = 0.95$ km) and 121 ($y = 1.66$ km) in Figure 17c. Figures 15d and 17d illustrate the total wavefield formed by

Profile I

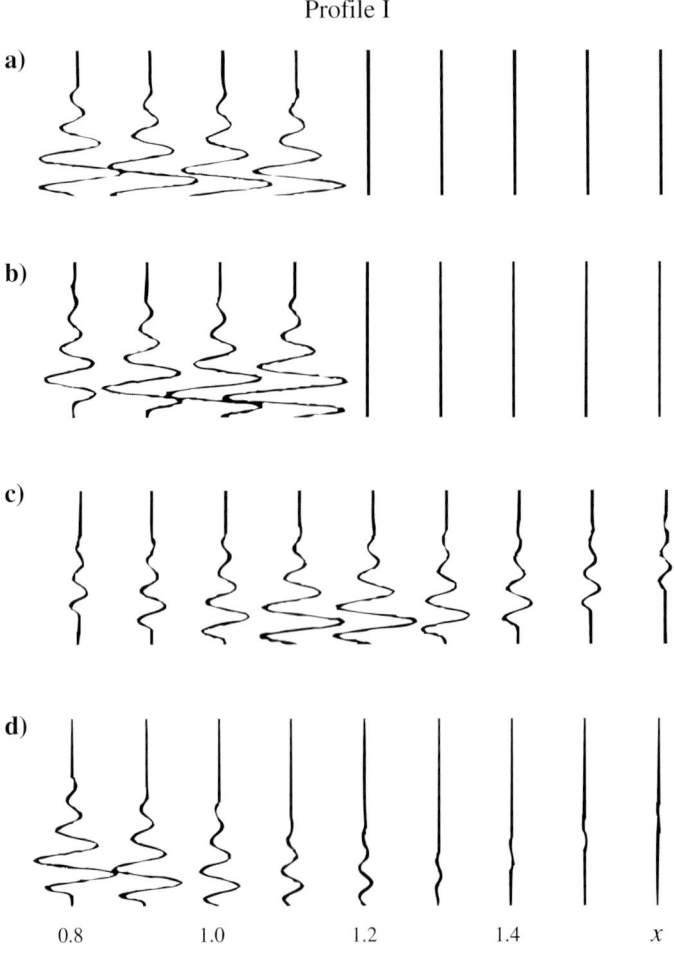

Figure 16. Details from theoretical seismograms in Figure 14: (a) reflected wave, (b) edge waves (scaled by a factor of 1.5), (c) tip waves (scaled by a factor of 7.5), and (d) total field (detail from Figure 12, Klem-Musatov and Aizenberg [1985]). © 1985, *Journal of Geophysics*. Used with kind permission of Springer Science and Business Media.

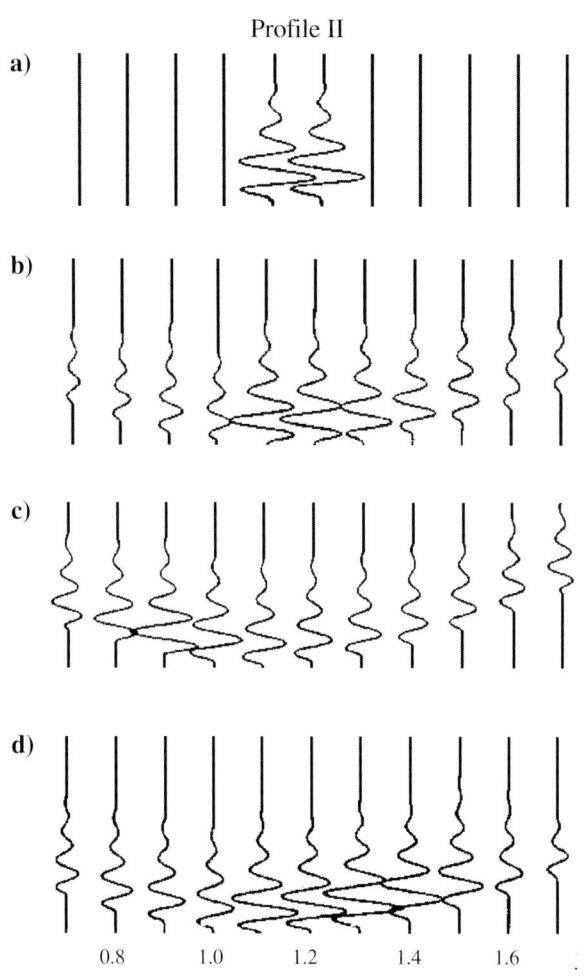

Profile II

a)

b)

c)

d)

0.8   1.0   1.2   1.4   1.6

Figure 17. Details from theoretical seismograms in Figure 15: (a) reflected wave (scaled by a factor of 0.5), (b) edge waves (scaled by a factor of 1.66), (c) tip waves (scaled by a factor of 4), and (d) total field (detail from Figure 13, Klem-Musatov and Aizenberg [1985]). © 1985, *Journal of Geophysics.* Used with kind permission of Springer Science and Business Media.

the interference of the reflected wave, two edge waves, and two tip waves. The total wavefield is regular everywhere.

Diffraction on a pyramid

Figure 18 depicts the reflecting interface and the observation system. The total wavefield can be written as

$$f = \sum_{m=1}^{3}\left[f_m + \sum_{n=1}^{2}(f_{mn} + f_{mn1}) \right],$$

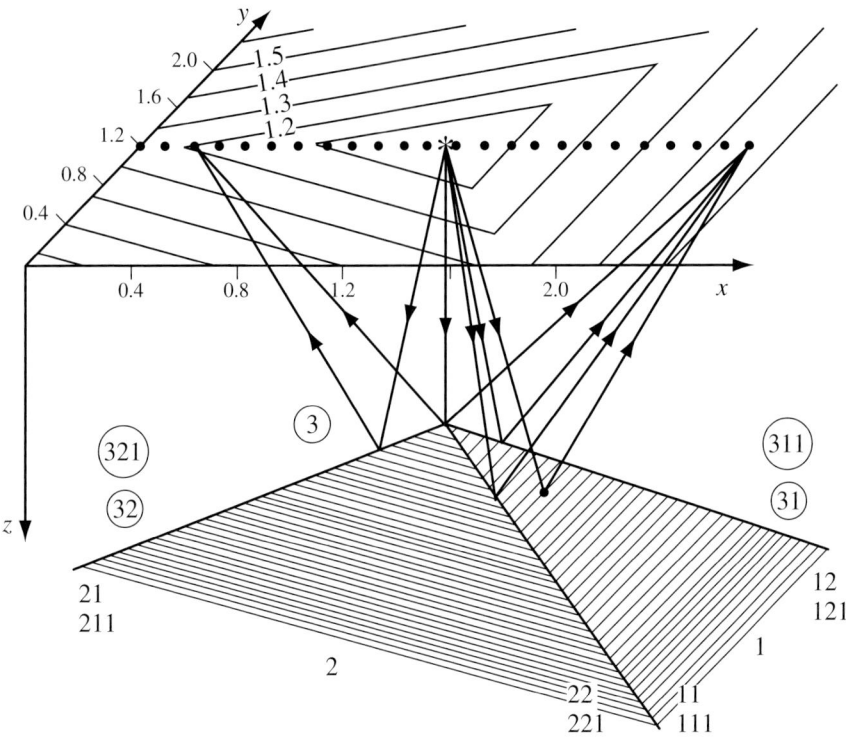

Figure 18. Model of a pyramid. The reflecting interface is given by depth lines on the plane of observation (from Figure 17, Klem-Musatov and Aizenberg [1985]). © 1985, *Journal of Geophysics*. Used with kind permission of Springer Science and Business Media.

where *m* is the index of a face, *n* is the index of the edge of the *m*th face, and f_{mn1} is a corresponding tip wave. All indices are illustrated in Figure 18 (indices connected with the invisible face are given in circles). Figure 19 illustrates wavefields scattered by the interface (Klem-Musatov and Aizenberg, 1985). Figure 20 enlarges the highlighted panels shown in Figure 19.

The observation profile is in the primary shadow zones of waves f_2 and f_3. The primary illuminated zone of wave f_1 can be seen in Figure 19. Figure 19b depicts the secondary illuminated zones of edge waves. Panels 1b through 1d in Figure 20 illustrate smoothing of the discontinuity at the secondary shadow boundary of waves f_{21} and f_{32}. Panel 1c in Figure 20 illustrates the phase inversion of corresponding tip waves. Panels 2b through 2d in Figure 20 demonstrate the smoothing of discontinuity at the secondary shadow boundary of waves f_{11}, f_{12}, f_{22}, and f_{31} by the tip

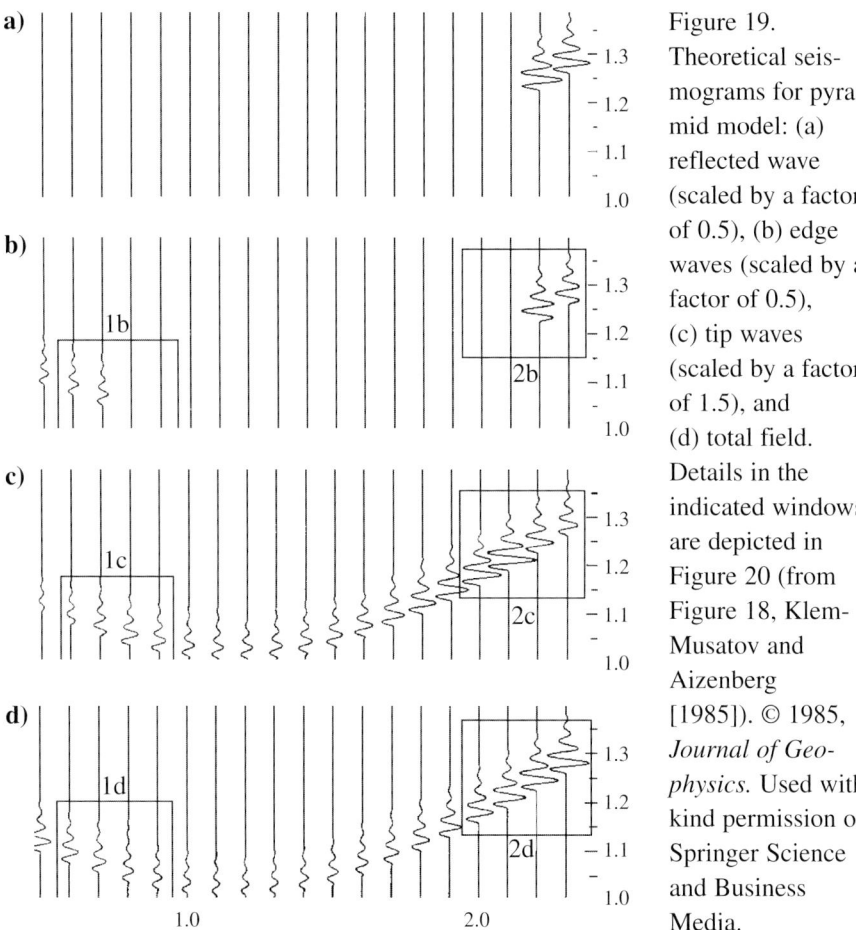

Figure 19. Theoretical seismograms for pyramid model: (a) reflected wave (scaled by a factor of 0.5), (b) edge waves (scaled by a factor of 0.5), (c) tip waves (scaled by a factor of 1.5), and (d) total field. Details in the indicated windows are depicted in Figure 20 (from Figure 18, Klem-Musatov and Aizenberg [1985]). © 1985, *Journal of Geophysics.* Used with kind permission of Springer Science and Business Media.

waves. Notice that all tip waves have the same eikonal and form a common diffracted wave, scattering from the vertex of the interface.

Diffraction of the transmitted wave on the dome (after Luneva and Kharlamov, 1990)

In a two-layer model, a plane longitudinal wave propagates from the lower medium into the upper. The interface consists of a flat boundary with a gentle dome in the form of a circle. The height of the dome and width of its base are λ and 6λ, respectively, where λ is the wavelength in the upper medium. The observation profile is located in the upper medium at a distance 3λ from the plane portion of the interface. Theoretical seismograms

Figure 20. Details from the theoretical seismograms in Figure 19: (b) edge waves (scaled by a factor of 0.5), (c) tip waves (scaled by a factor of 1.5), and (d) total field (detail from Figure 18, Klem-Musatov and Aizenberg [1985]). © 1985, *Journal of Geophysics.* Used with kind permission of Springer Science and Business Media.

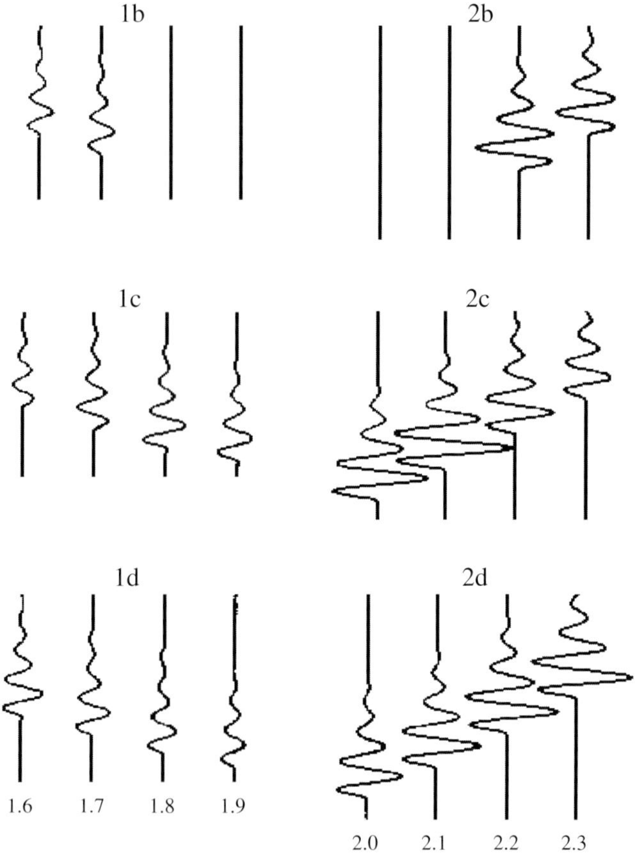

are calculated by the edge-wave single superposition method. The physical experiment is carried out on a 2D model composed of aluminum and Plexiglas® sheets. Figures 21 and 22 show the x (horizontal) and z (vertical) components, respectively, of the theoretical and experimental seismograms and amplitude graphs.

Implementation in a 2D modeling package

The simplicity of the boundary-layer approximation makes it easy to use in seismic modeling and prospecting, as shown in several cited papers. This approximation also is used for diffraction modeling in a 2D modeling software package developed for Norsk Hydro and the University of Bergen (Pajchel et al., 1987). Here, we demonstrate its application in simple 2D models and in more realistic situations that involve complex geologic struc-

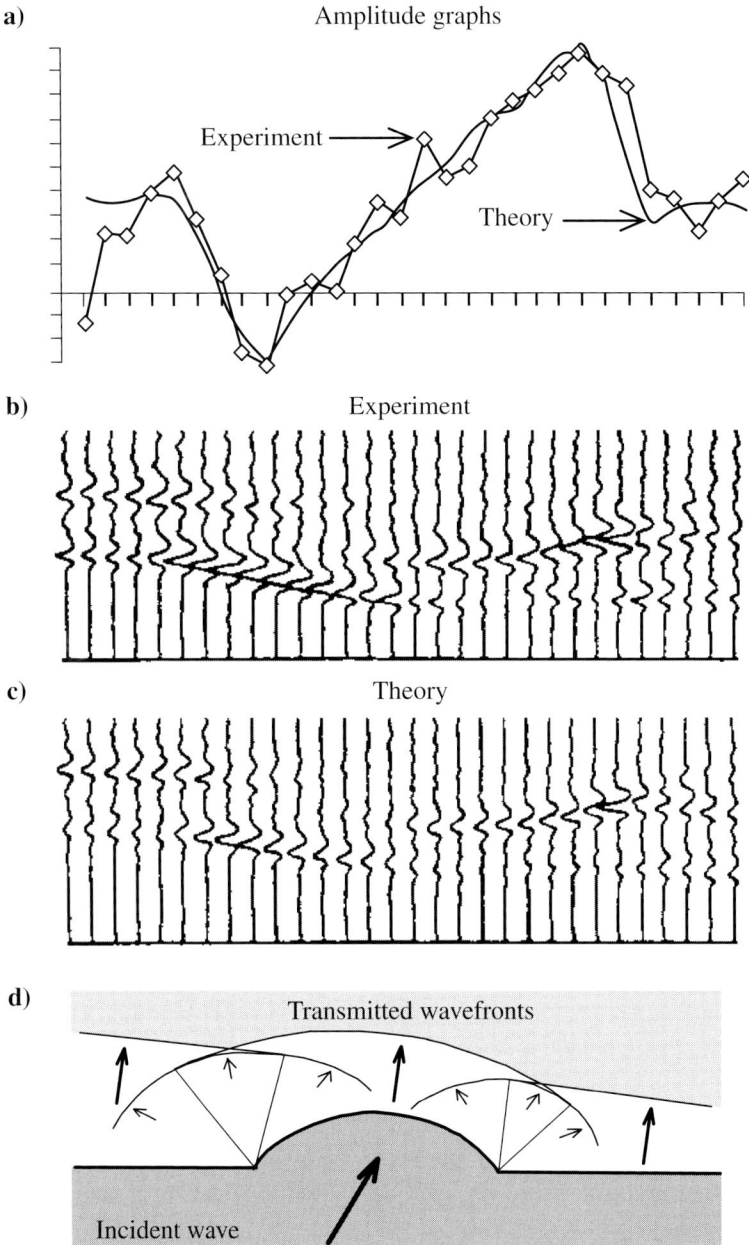

Figure 21. (a) Amplitude graphs, (b) experimental seismogram, and (c) theoretical seismogram of the longitudinal wavefield *x* component for (d) transmission of the plane wave through the flat dome. Domains of transmitted waves are shown in white (from Luneva and Kharlamov, 1990).

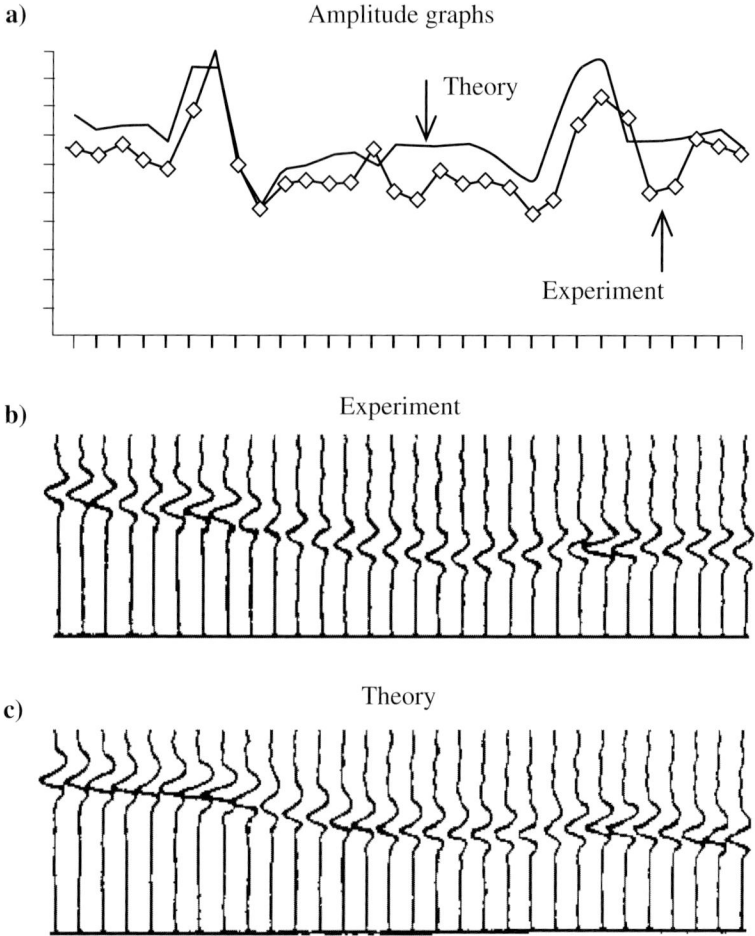

Figure 22. (a) Amplitude graphs, (b) experimental seismogram, and (c) theoretical seismogram of the longitudinal wavefield z component for transmission of the plane wave through the flat dome (from Luneva and Kharlamov, 1990).

tures. Klaeschen et al. (1994) has described a similar 2D package (also based on combining the standard ray method and edge-wave approach).

Package organization

The 2D modeling software package (Pajchel et al., 1987) is designed to serve a broad range of seismologic applications, ranging from crustal and earthquake seismology to vertical-seismic-profile (VSP) exploration for hydrocarbons. Figure 23 portrays an overview of the software package and the functional relationship between the individual programs modules. As indicated in the flow diagram, the programs are grouped into five categories according to mode of operation.

Model design

The first group of programs is for the input of the geologic model. The program DIGITIZE transfers the model geometry from a digitizing table. MODGEN is the model generator, which contains options for geometry editing, input of seismic velocities, source and receiver location, and so forth, and the input of various other parameters relevant for subsequent ray

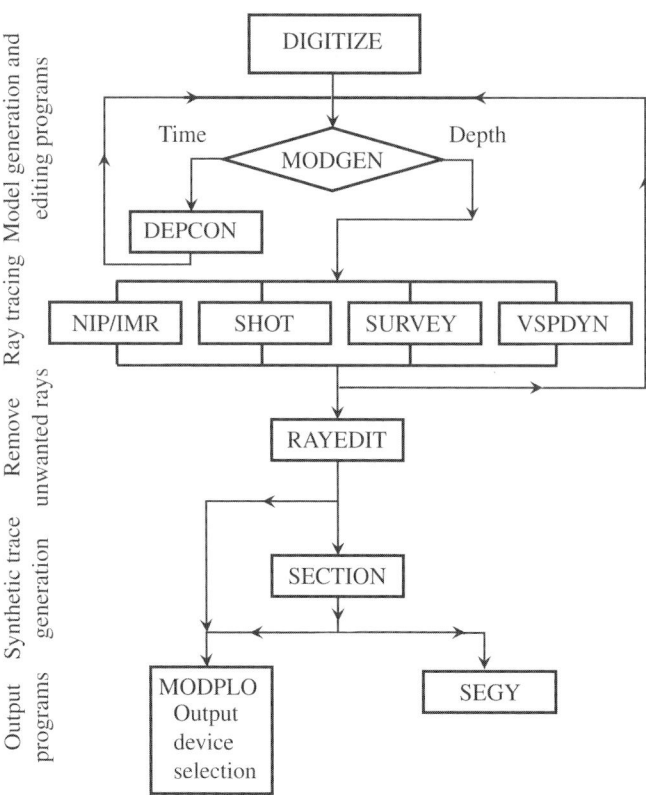

Figure 23. Structural overview of modeling programs, describing the interconnection among model-building modules (DIGITIZE, MODGEN, and DEPCON), wave-modeling modules (NIP/IMR, SHOT, SURVEY, and VSPDYN), and synthetic seismogram generation (SECTION) with output modules (MODPLO and SEGY) (from Pajchel et al., 1987).

tracing. If the model geometry is given in the form of a time section, conversion to depth is made first by DEPCON before proceeding to one of the ray-tracing programs. Geometric representation is based on the interface logic concept, implying that each interface must be defined throughout the model frame.

Because ray tracing requires the interface to be defined and continuous at any point between the digitized points, we apply piecewise cubic-spline interpolation. We attach a code to the corner point to preserve sharp corners or edges, which cancels smoothing locally. In ray-tracing programs, this corner-point code also signals that edge diffraction can be computed.

Ray tracing

Five ray-tracing programs are included to handle various types of seismologic applications. The normal-incidence profiling (NIP) program simulates a common-source-receiver seismic survey by tracing normal rays from a given subsurface reflector to the surface. IMR, a modification of NIP, is made to handle image rays that are traced into the subsurface perpendicularly from surface datum. SHOT is a general-purpose ray-tracing program that can compute rays from a source (arbitrarily located in the model) to a receiver array, which is confined to a horizontal line. The most common seismology application is simulation of a single shot in conventional seismic exploration, where source and receivers are placed at the surface of the earth. However, because the source can be positioned anywhere in the model, SHOT also is suited well for studies of local earthquakes where the curvature of the earth can be neglected.

SURVEY is an automated version of SHOT — source and receivers can be moved stepwise to simulate a conventional seismic survey. For each step, a new synthetic shot gather will be generated. In simulation of a conventional surface seismic survey, steps are horizontal. For the special type of well-seismic survey known as walkaway VSP, the step in the receiver position can be vertical (or along some deviated track), although the source array remains fixed at the surface. In the latter configuration, it is convenient (by applying the principle of reciprocity) to interchange the source and receiver positions, i.e., the source is positioned in the borehole, and receivers are at the surface.

VSP is designed to handle various problems involved in commonly used well-seismic configuration where the source is located at a fixed position near the surface. Here, the string of receivers in the wellbore intersects several layers of the model, implying dramatic changes in the ray code between closely neighboring rays. This program is designed to handle such problems and to admit separate tracing of the upgoing and downgoing wavefields. Options for single or multiple ray intersection with the wellbore also are included in VSP. They are needed sometimes if the model contains a complex geometry or velocity distribution.

Diffraction computation

The asymptotic ray-theory (ART) part of the package is based essentially on Červený et al. (1977). Head waves are computed according to the formulas presented in Červený and Ravindra (1971). Therefore, the diffraction computations performed by this package include both reflection diffraction and head-wave diffraction. The two concepts are displayed in Figure 24.

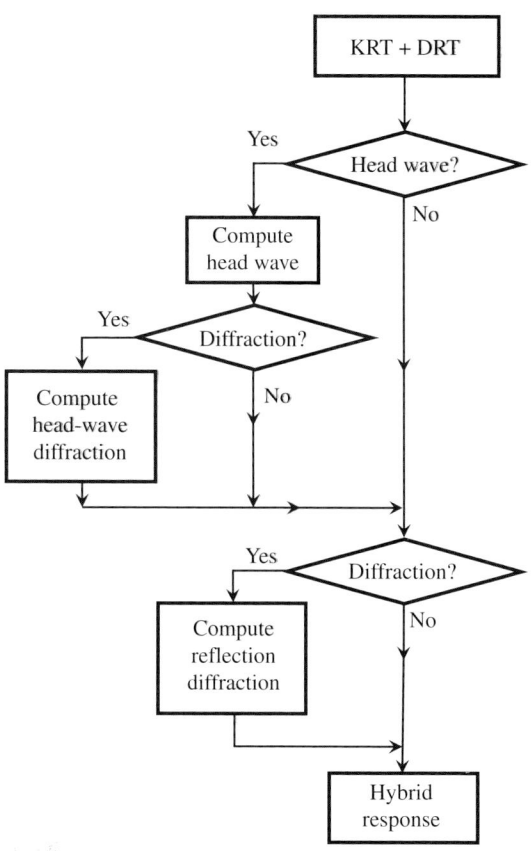

Figure 24. Computational procedure for head waves and diffractions. Based on results of kinematic (KRT) and dynamic ray tracing (DRT), tests are made to determine whether any rays approach critical angle and head waves are generated at the interface (the answer is yes). If that interface has a point of discontinuity (yes), the head-wave diffraction can be computed. If no head waves are detected and the interface contains a point of discontinuity, then only reflection diffractions are computed (from Pajchel et al., 1987).

Three conditions must be fulfilled to initialize the diffraction computation at a particular point. First, the point must be a corner point of a model interface. Second, if the first condition is satisfied, an additional code must be attached to the point that identifies whether diffractions are desired and what type of diffraction should be computed. Third, if the first two conditions are met, the neighborhood of the point must be illuminated by the appropriate type of rays to determine shadow boundaries.

MODGEN and the individual ray-tracing programs include a graphic editor for the input and editing of diffraction points. Thus, diffractions can be specified in advance and modified in the ray-tracing programs. Diffraction computations in a 2D model imply that only one boundary ray must be defined for each side of the corner point. The reflection-diffraction boundary ray is determined by the reflected ray at the edge. The head-wave boundary ray is determined by the angle of the emerging ray equal to the incident angle of the critical ray.

The ray-tracing programs are designed to handle the boundary ray at the edge. The head-wave boundary ray is determined by the angle of the emerging ray equal to the incident angle of the critical ray.

A combined two-point tracing and shooting technique is applied. Two-point tracing is used to find the raypath between the source and the corner point. The shadow boundary for each side of the corner is determined by shooting rays from the edge using the initial directions. Once travel path and amplitude of the boundary ray are determined, diffractions are evaluated by shooting a bundle of rays, centered on the boundary ray, from the edge point to the receiver interface. A diffraction coefficient is attached to each of these rays (essentially ART rays), using a specified wavelet center frequency ω.

Figure 24 is a flow diagram illustrating the computational procedure. When the area around the diffraction point has been illuminated by ART rays, the kinematic (KRT) and dynamic (DRT) parameters of the boundary ray are determined, and the diffractions are computed automatically.

Computation of reflection diffractions has been demonstrated already in the simple wedge model formed by two intersecting reflectors, portrayed in Figure 5. These diagrams demonstrate explicitly one of the underlying assumptions of edge-wave theory: The resulting superposition of reflections and diffractions forms a smooth response. This also demonstrates clearly the ability of the approach of the edge wave to preserve the advantage of ray-based modeling, separating the wavefield into individual arrivals.

Vertical seismic profiling

In vertical seismic profiling (VSP), a string of densely spaced receivers is located along a vertical or deviated borehole. The source in most VSP configurations is placed at the earth's surface. Because the string of receivers can intersect many layers, the tracing of rays in VSP simulations is characterized by a frequent change of ray code. Therefore, an automatic ray-code generator is included in the package. This is particularly useful when a large number of diffracted rays must be traced to a string of receivers in a multilayer VSP simulation. Strong diffractions of P- or S-waves often characterize real VSP data because the geophones can be close to geologic features that cause diffractions. Modern processing techniques and the latest VSP tools, equipped with three-component geophones and a gyrocompass, make it possible to record and enhance a larger variety of seismic events than ever before.

Rather than classical brute-force enhancement of P-wave reflections, which implies rejection of other information in the data, the modern approach is to exploit information contained in the mode converted and the diffraction events. Therefore, carefully performed modeling studies are increasingly more important in designing a VSP survey, both in the processing of data and the interpretation of results. Figures 25 through 28,

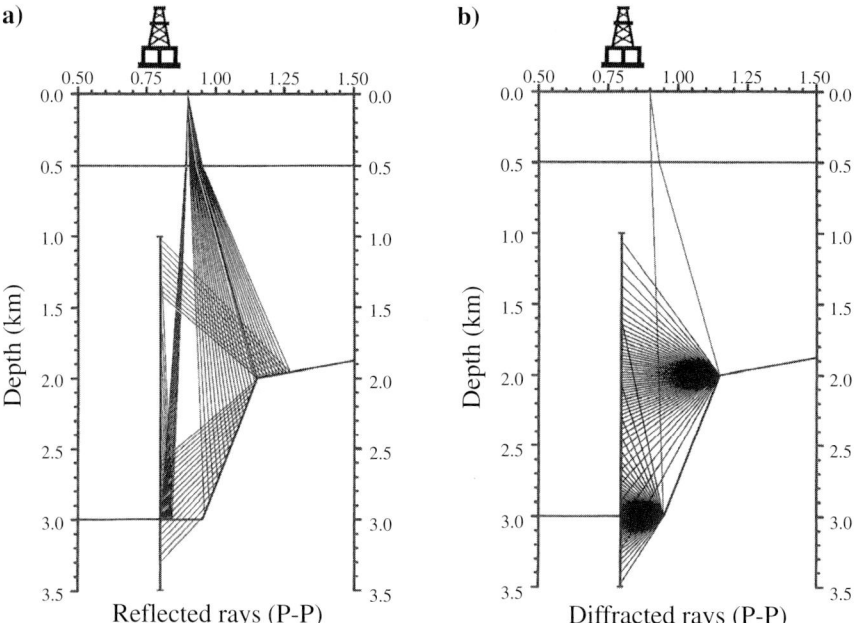

Figure 25. Ray diagrams for (a) reflections and (b) diffractions in a rig-source (zero-offset) vertical seismic profile. The model is offshore with seabed at 0.5 km. The target reflector is formed like a staircase (inclined fault), stepping from 3 km to about 2 km in depth. Model interfaces are shown in solid lines (from Pajchel et al., 1989).

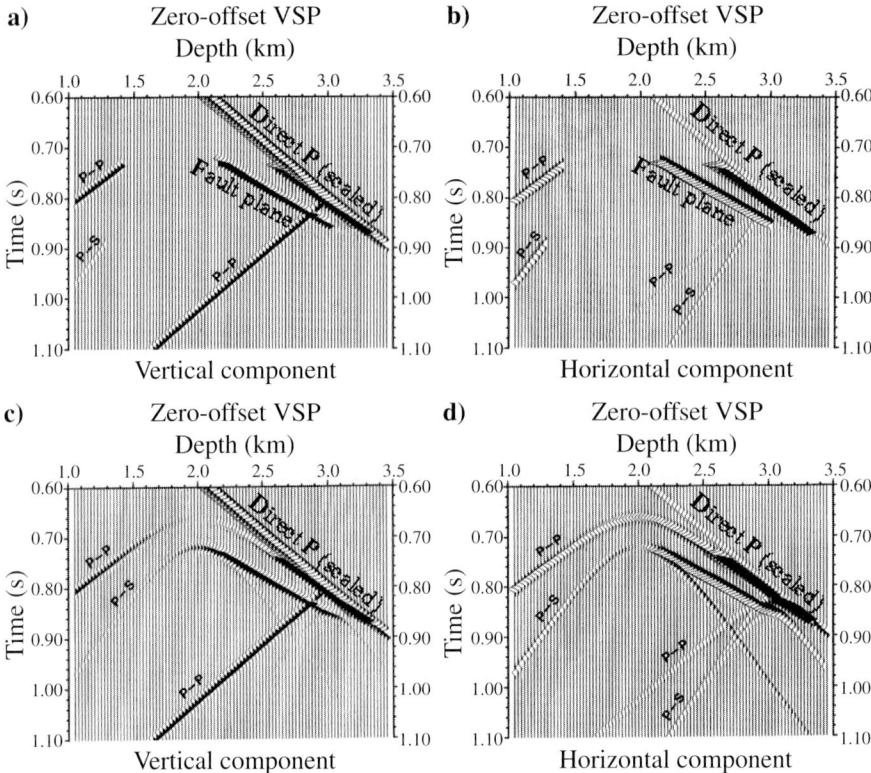

Figure 26. (a) Vertical component for zero-offset VSPs without diffractions, (b) horizontal component for zero-offset VSPs without diffractions, (c) vertical component for zero-offset VSPs with edge waves superimposed, and (d) horizontal component for zero-offset VSPs with edge waves superimposed. The geologic model is the same one depicted in Figure 25 (from Pajchel et al., 1989).

using a simple model with a vertical well, demonstrate that the diffraction response constitutes a significant part of the seismic picture. Without diffractions, seismograms appear to be far from realistic, consisting of isolated, discontinuous events. By applying the combined ART and edge-wave technique, it is possible to synthesize a complex combination of mode conversions and diffractions while retaining the important advantage of ray modeling — all generated events are separable and thus easy to identify.

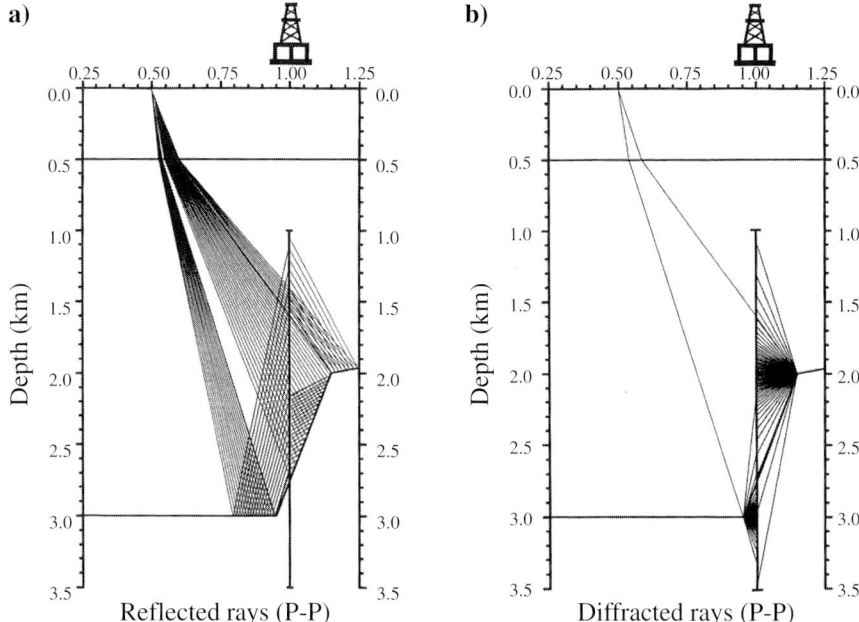

Figure 27. Ray diagrams for (a) reflections and (b) diffractions in an offset-source VSP (from Pajchel et al., 1989).

Seismic survey modeling and processing of synthetic data

In this example, we have synthesized a seismic survey with approximately 200 individual shot panels. The main objective of this study (Frøyland et al., 1988) is to evaluate possibilities and limitations in the conventional seismic method for a given geologic situation, depicted in Figure 29. The target consists of a series of rotated fault blocks, modeled as two sequences of broken reflectors. Detection of the small dipping reflectors at great depths is supposed to be near the resolution limit of the seismic method.

To evaluate the effects on the data that can be introduced by various steps in data processing, i.e., CDP stacking, migration, and so forth, the synthetic shot gathers are transferred to a seismic data-processing system. Test results are shown in Figures 30 through 32.

Figure 30 displays the lower part of a synthetic shot panel, corresponding to that of the rays in Figure 29. Reflections from the lower four reflectors and diffractions from fault edges are shown. Reflections from the

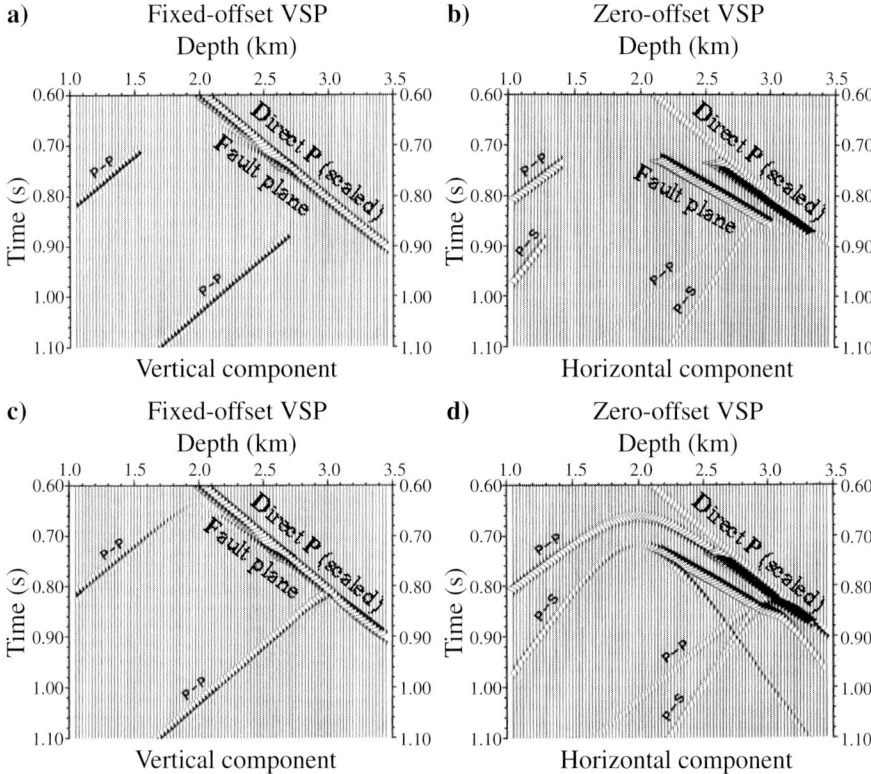

Figure 28. (a) Vertical component for offset-source VSPs without diffractions, (b) horizontal component for offset-source VSPs without diffractions, (c) vertical component for offset-source VSPs (top) with edge waves superimposed, and (d) horizontal component for offset-source VSPs with edge waves superimposed (from Pajchel et al., 1989).

major fault plane in the central part of the model are of particular interest. With the present source-and-receiver distribution, fault-plane reflections are recorded only by receivers near the source. The fault plane can be identified in the shot gather as one of the events dipping opposite to normal spread dip. Diffractions attached to the fault-plane reflection itself and diffractions entering from edges located on both sides have been included.

Because we have included edge diffractions, it is now possible to study the effects of seismic migration in a realistic way. The power of prestack depth migration is realized by comparing the depth-migrated gather displayed in Figure 30 with the model in Figure 29. Based on a single shot, the

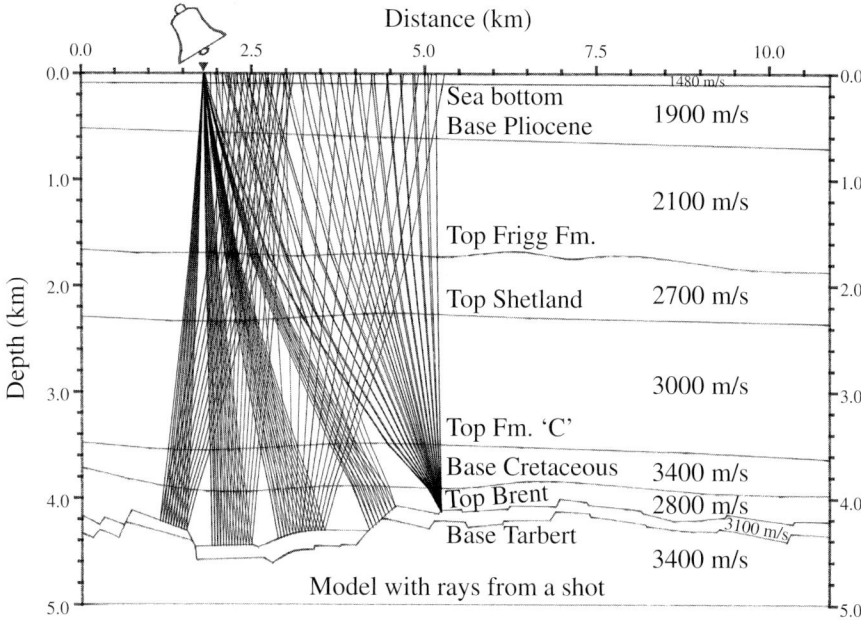

Figure 29. Geophysical model subjected to survey modeling study (from Frøyland et al., 1988).

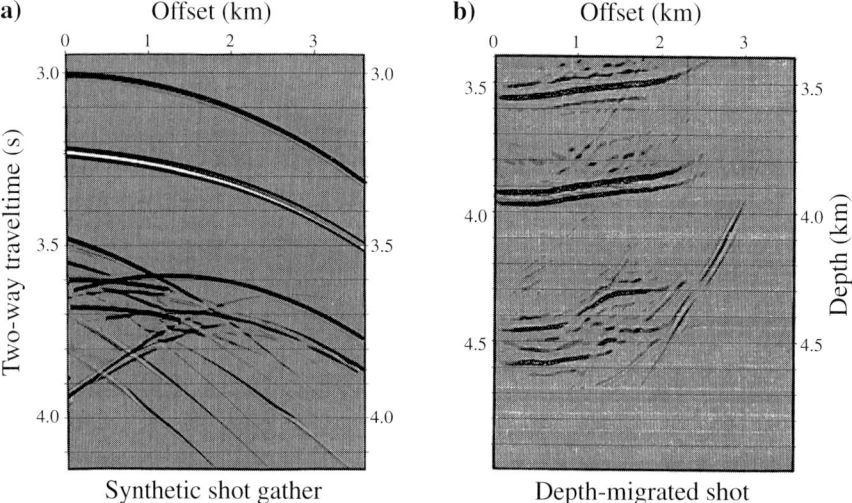

Figure 30. (a) Synthetic shot gather and (b) its depth-migrated version. The corresponding model and rays are shown in Figure 29 (from Frøyland et al., 1988).

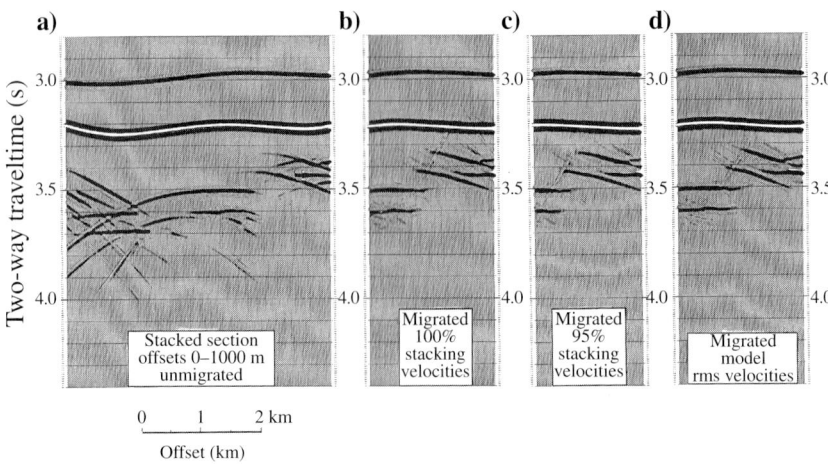

Figure 31. (a) Unmigrated CDP-stacked section and its migrated versions with velocity models using (b) 100% and (c) 95% stacking velocity and (d) rms velocity (from Frøyland et al., 1988).

steeply dipping fault plane and the adjacent reflectors are reasonably well mapped into their correct positions.

Depth migration of each shot gather is expensive in terms of computer time. Therefore, the latter procedure is applied rarely if target features can be enhanced properly by using the less computationally intensive poststack migration. As indicated by results displayed in Figure 31, a poststack migration will be adequate in the present case. When a reasonably correct velocity model is applied in the migration, the fault plane is well mapped into its correct position. Another important conclusion from this study is that unless the survey is extended beyond the left frame of the model (Figure 29), only the near-source part of the shot panels (Figure 30) contains reflections from the fault. Therefore, including additional channels in the stack will result only in the addition of noise and thus will reduce the stacking response of the fault plane. Eliminating the far-offset part of the data clearly improves the result.

Normal-incidence profiling (NIP)

NIP ray tracing is an adequate approach for many problems in conventional surface-seismic exploration, e.g., in the interpretation of stacked sections, testing related to poststack migration, and so forth. We apply the edge-wave concept in the NIP program for diffraction modeling also.

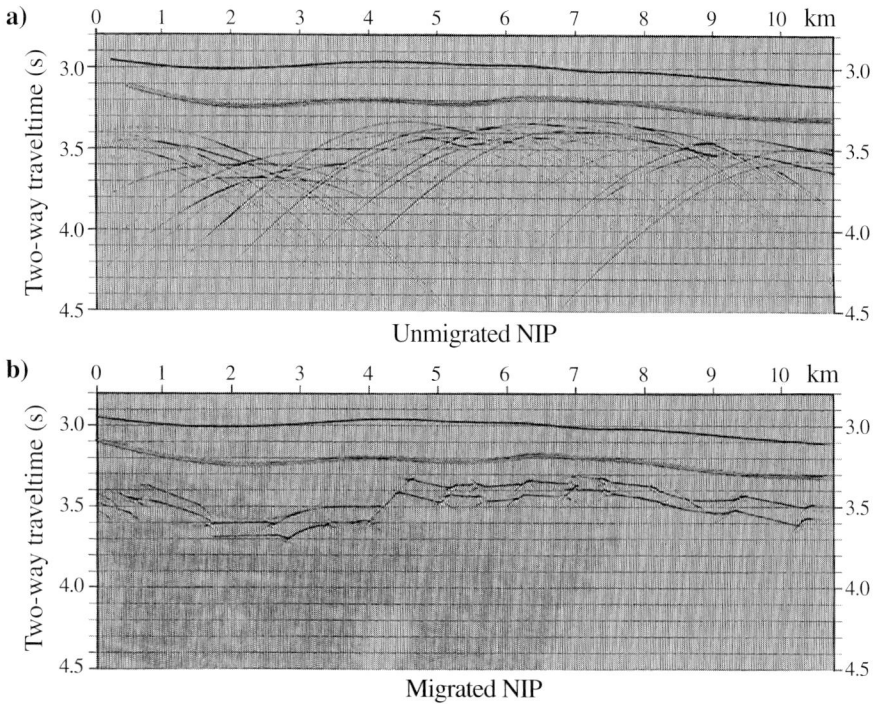

Figure 32. Synthetic data generated by the NIP program illustrating (a) a response dominated by diffractions and (b) its migrated version, which reproduce the model geometry (Figure 29) (from Frøyland et al., 1988).

Figure 32 illustrates an example of its application. Again, we use the model displayed in Figure 29 and focus our attention on the deep fault blocks. The unmigrated NIP synthetic section is dominated by diffracted events. Individual reflections are difficult to identify. However, after applying a straightforward *fk*-migration procedure from the seismic processing package, the fault blocks of the model are clearly visible.

The tip-wave superposition method

The theory of edge waves fails where the ray-theory field changes rapidly, e.g., in caustic zones. These shortcomings are eliminated in the tip-wave superposition method (TWS), which provides realistic seismograms for models with curved boundaries and faults. The TWS also applies to 3D models, but it is restricted here to two layers separated by a single interface. Implementation for multilayer models is under development (Ayzenberg et al., 2007b). The initial idea of this method is based on consideration of the mechanism of scattering waves by a polyhedral interface (Klem-Musatov et al., 1982). We notice that this kind of interface behaves like a smooth interface when polyhedron faces are sufficiently small and (unlike in Klem-Musatov et al., 1982) when they scatter only from the vertices of the interface, contributing to the reflected (transmitted) wavefield.

The TWS approach can be related to the Kirchhoff integral solution of the wave equation. Here, integration over the reflecting interface, divided into a triangular grid, sums the contribution to the wavefield from each plane triangle caused by a beam of rays. The beam includes one geometric ray, three edge-wave rays, and three tip-wave rays. Rays and traveltimes for each geometric element on a given geologic interface are implicit in the computation and thus are available.

Algorithm

Let two homogeneous elastic media be separated by an infinite smooth interface S. A concentrated source generates a nonstationary elastic wave in the upper half space. The problem of finding the scattered wavefield can be reduced to the stationary case by using a time-frequency Fourier transform. Then the corresponding mathematical problem can be reduced to the solution of a system of wave equations for elastic potentials satisfying corresponding conditions at the interface. We describe how the individual scalar components of the solution can be described in a high-frequency approximation.

Any scalar component f of the reflected (or transmitted) wavefield can be represented by Kirchhoff's integral,

$$f = \int_S \left(\overline{f}G_n - \overline{f}_n G\right)dS. \tag{1}$$

where $\overline{f}$ and $\overline{f}_n$ are the boundary values of f and its normal derivative on S, respectively. G and G_n are the Green's function and its normal derivative on S, respectively.

If the surface S is approximated using triangulation (Figure 33), the wavefield in equation 1 can be written as a superposition of contributions from the elementary triangles S_m:

$$f = \sum_m \Delta f_m, \quad \Delta f_m = \int_{S_m} \left(\overline{f}G_n - \overline{f}_n G\right)dS. \tag{2}$$

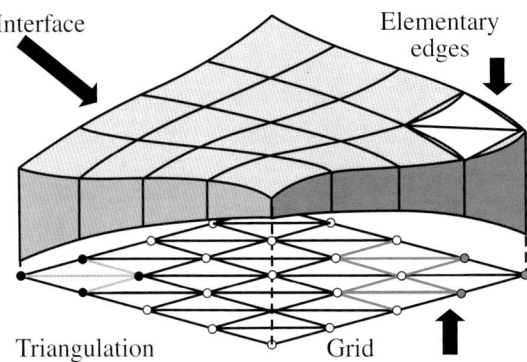

Figure 33. The triangulated interface with its projection on the gridded structural map.

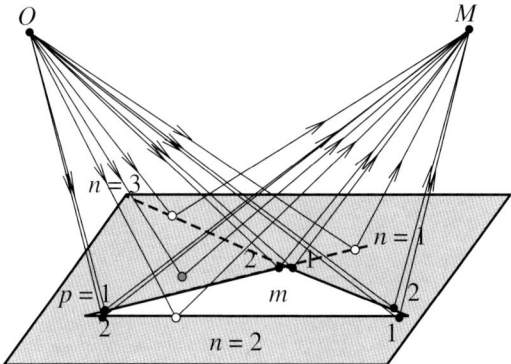

Figure 34. Scheme of contributions from the individual triangle. Diffracted waves are described in the boundary-layer approximation that is valid in the vicinity of the shadow boundaries.

In the high-frequency approximation, the contribution from an individual triangle can be derived by the theory of edge waves or by direct asymptotic analysis of the integrals Δf_m,

$$\Delta f_m = \delta_m f_m + \sum_{n=1}^{3} \left(\delta_{mn} f_{mn} + \sum_{p=1}^{2} f_{mnp} \right), \qquad (3)$$

where f_m is a wave reflected from the mth triangle S_m, f_{mn} is a wave diffracted from the mnth edge of the triangle, and f_{mnp} is a wave scattered from the mnpth tip of the edge (Figure 34).

The edges of the elementary triangles define primary shadow zones. Each primary mnth shadow boundary and its analytic continuation into space divide space into primary illuminated Ω_{mn}^- and primary shadow Ω_{mn}^+ zones. Thus, in equation 3, we have

$$\delta_m = 1 \quad \text{when} \quad M \in \Omega_{mn}^-,$$
$$\delta_m = 0 \quad \text{when} \quad M \in \Omega_{mn}^+, \qquad (4)$$

where M is the current point of space. The break points of the edges cause the secondary shadow zones in the edge-diffracted waves. Each secondary *mnp*th shadow boundary and its analytic continuation into space divide space into secondary illuminated Ω_{mnp}^{-} and secondary shadow Ω_{mnp}^{+} zones. Thus, in equation 3, we have

$$\delta_{mn} = 1 \quad \text{when} \quad M \in \Omega_{mnp}^{-}, \quad \delta_{mn} = 0 \quad \text{when} \quad M \in \Omega_{mnp}^{+}. \quad (5)$$

The reflected (or transmitted) wave is described in the framework of asymptotic ray theory,

$$f_m = \Phi_m \exp\left(i\omega\tau_m\right), \quad (6)$$

where Φ_m and τ_m are the amplitude and eikonal, respectively, and ω is the angular frequency.

Diffracted waves are described in the boundary-layer approximation, valid in the vicinity of the shadow boundaries,

$$f_{mn} = s_{mn}W\left(w_{mn}\right)\Phi_m \exp\left(i\omega\tau_{mn}\right) \quad \text{and}$$
$$f_{mnp} = s_{mnp}H\left(\rho_{mnp}, \zeta_{mnp}\right)\Phi_m \exp\left(i\omega\tau_{mnp}\right), \quad (7)$$

where $W(w)$ and $H(\rho, \zeta)$ are the edge and tip diffraction functions.

The arguments of the diffraction functions are

$$w_{mn} = \sqrt{2\omega\left(\tau_{mn} - \tau_m\right)/\pi}, \quad \rho_{mnp} = \sqrt{2\omega\left(\tau_{mnp} - \tau_m\right)/\pi} \quad \text{and}$$
$$\zeta_{mnp} = \arcsin\sqrt{\left(\tau_{mnp} - \tau_{mn}\right)/\left(\tau_{mnp} - \tau_m\right)}, \quad (8)$$

where τ_{mn} and τ_{mnp} are the eikonals of the edge and tip waves, respectively. In these formulas, functions $\Phi_m(x, y, z)$, $\tau_m(x, y, z)$, and $\tau_{mn}(x, y, z)$ are continued analytically from the corresponding primary and secondary illuminated zones into the shadow zones.

Unit-step functions describe the discontinuities of diffracted waves at shadow boundaries:

$$s_{mn} = \begin{cases} +1 & \text{when} \quad M \in \Omega_{mn}^{+}, \\ -1 & \text{when} \quad M \in \Omega_{mn}^{-}, \end{cases} \quad s_{mnp} = \begin{cases} +1 & \text{when} \quad M \in \Omega_{mnp}^{+}, \\ -1 & \text{when} \quad M \in \Omega_{mnp}^{-}. \end{cases}$$

$$(9)$$

Because the interface is approximated by a set of plane triangles, all terms in equation 3 are bounded in the caustic zones caused by curvature of the interface. If the grid size is sufficiently small, only tip waves contribute to the sum in equation 2. That is why we call this approach the tip-wave superposition method (TWS).

Expressions 2 and 3 allow us to visualize the seismic response of the interface by highlighting edge and tip projections on a gridded map of the interface and showing absolute amplitude values with a standard color scale. We call such a map the diffraction-response pattern.

Implementation and software overview

A collection of FORTRAN 77 routines constitutes the prototype software package (Aizenberg et al., 1992a, 1992b) used in the following experiments. A driver program PROFILE (Figure 35) calls 18 subroutines. In addition, an auxiliary 3D model generator (Pajchel et al., 1988) was used to prepare the gridded surface. Subroutine BOUNDARY reads the 2D array of

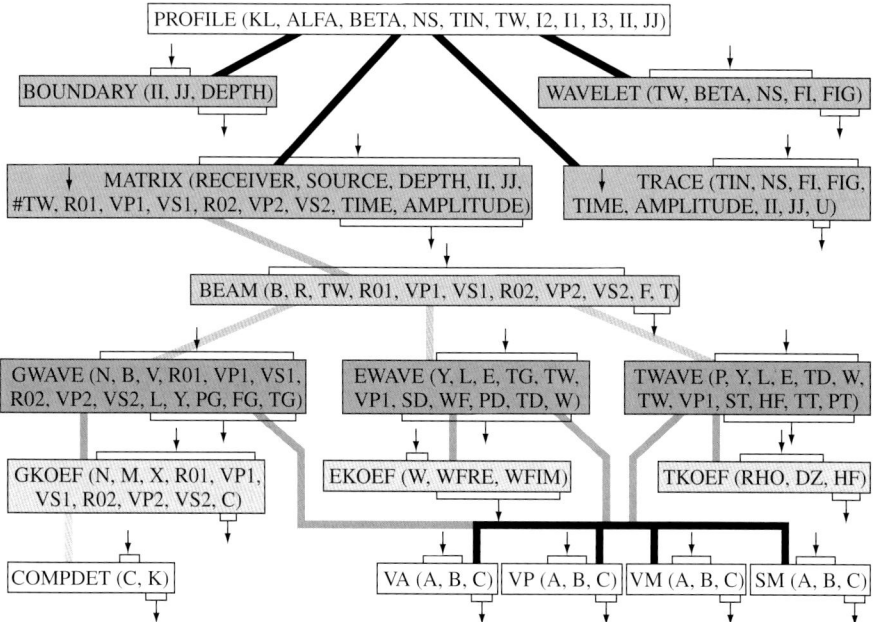

Figure 35. Program structure for the tip-wave superposition method (TWS). See text for details (from Aizenberg et al., 1992b).

depth values and reorganizes the rectangular grid into triangles as in Figure 33. There are provisions for a nonuniform grid.

MATRIX is a driver routine for subroutine BEAM, which computes seven waves (one geometric wave, three edge waves, and three tip waves), according to equations 6 and 7, by calls to GWAVE, EWAVE, and TWAVE, respectively. GWAVE, EWAVE, and TWAVE call GKOEF, EKOEF, and TKOEF to evaluate the kinematic and dynamic parameters. GKOEF calls COMPDET, which evaluates the determinant of a complex-valued matrix. EWAVE calls EKOEF, which computes the frequency-dependent edge-wave diffraction coefficient. TWAVE calls TKOEF, which computes the tip-wave diffraction coefficient based on equation 7. Special function $H(\rho, \zeta)$ is interpolated from a table. The four routines VA, VP, VM, and SM are modules for basic vector operations.

MATRIX outputs arrays containing arrival time and complex-valued amplitude vectors of constituent waves for the given source-receiver configuration. Subroutine WAVELET supplies the analytical Berlage signal and its Hilbert transform. Seismic traces are computed by the routine TRACE. An additional module, DECOMP, maps kinematic and dynamic results on the surfaces.

Experimental results

This experimental model (Aizenberg et al., 1992a, 1992b) is taken from a North Sea oil field, where the target is an interface with complex fault patterns (Figure 36). Model geometry is based on uniformly gridded structural maps. Only one interface is used, and it separates two homogeneous layers with the following properties:

$$\rho_1 = 2.5 \text{ g/cm}^3 \quad v_{P1} = 2.0 \text{ km/s} \quad v_{S1} = 1.5 \text{ km/s}, \quad \text{and}$$
$$\rho_2 = 2.6 \text{ g/cm}^3 \quad v_{P2} = 4.0 \text{ km/s} \quad v_{S2} = 3.0 \text{ km/s}.$$

In geologic terms, the model is a slice taken from an uplifted horst block, with steep bounding faults. Its top is intersected by several faults (Figures 36 through 38). A grid of 125×125 regular cells of size 25 m $\times$ 25 m was used. Only part of the grid (Figure 36) contains physical data, which are selected from the grid with a screening matrix. This matrix contains values 0 or 1 to flag data located outside or inside the physical part of the surface.

In this experiment, we select three seismic lines, as indicated in Figure 38. Line directions are chosen according to the original seismic survey, which

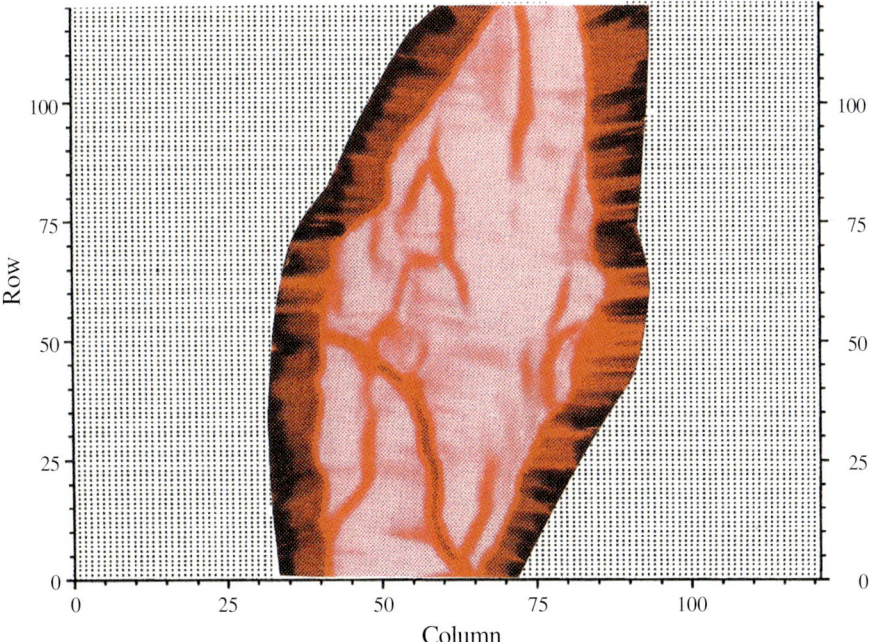

Figure 36. Modeling grid. The physical part of the surface is color-shaded. Faults are indicated by dark patterns (from Aizenberg et al., 1992b).

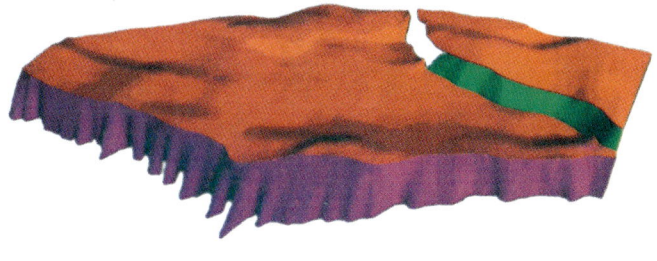

Figure 37. 3D perspective view of the model reflector as seen from the north-northwest (from Aizenberg et al., 1992b).

forms the background data for the model. We simulate a zero-offset survey with the source-receiver location at a depth of 5 m. The source is a point source of spherical P-waves. The synthetic section wavelet is the Berlage signal with the two alternative frequencies of 20 and 50 Hz.

We compute the three components of the displacement vector. The x-axis is directed along the profile line, the y-axis is perpendicular to the profile plane, and the z-axis is along the vertical. All wave sections are represented by the total P-wavefield and separately by individual sums for each wave type.

Figure 38. Detailed structural depth map of the test reflector showing the seismic lines used in this study. A, B, C, D, and E indicate the faults discussed in the text (from Aizenberg et al., 1992b).

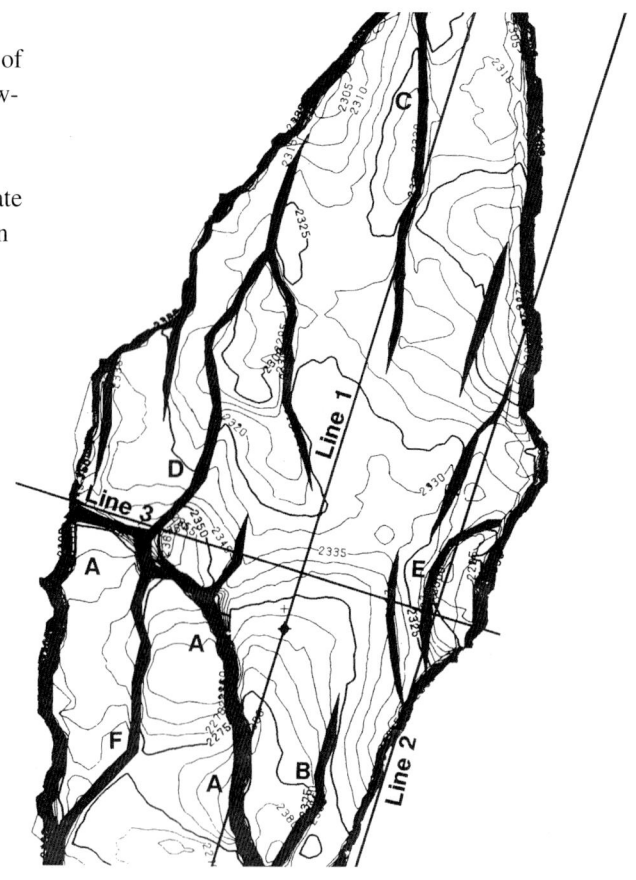

We map points on the interface that contribute to the recorded wavefield, and we call these points diffraction-response patterns (DRPs). For each source-receiver point, the entire surface is considered in the computation. However, we show only grid points that contribute with signal amplitude above a given threshold in the result. For each seismic line, results are organized in the following manner: Traces of the three components of the total wavefield are displayed first, followed by corresponding traces of the tip, edge, and reflected waves. All traces presented are normalized and colorcoded according to the standard seismic color scale.

Line 1

The total P-wave synthetics in Figure 39 are rather complex and rich in detail, but they reveal quite coherent events. In particular, the two horizontal components Σ_x and Σ_y are influenced strongly by diffractions and by events originating from locations offset far from the line. In the vertical component section Σ_z, the main signals from the upthrown and downthrown sides of fault A are clearly recognizable. Significant lateral-amplitude variations along

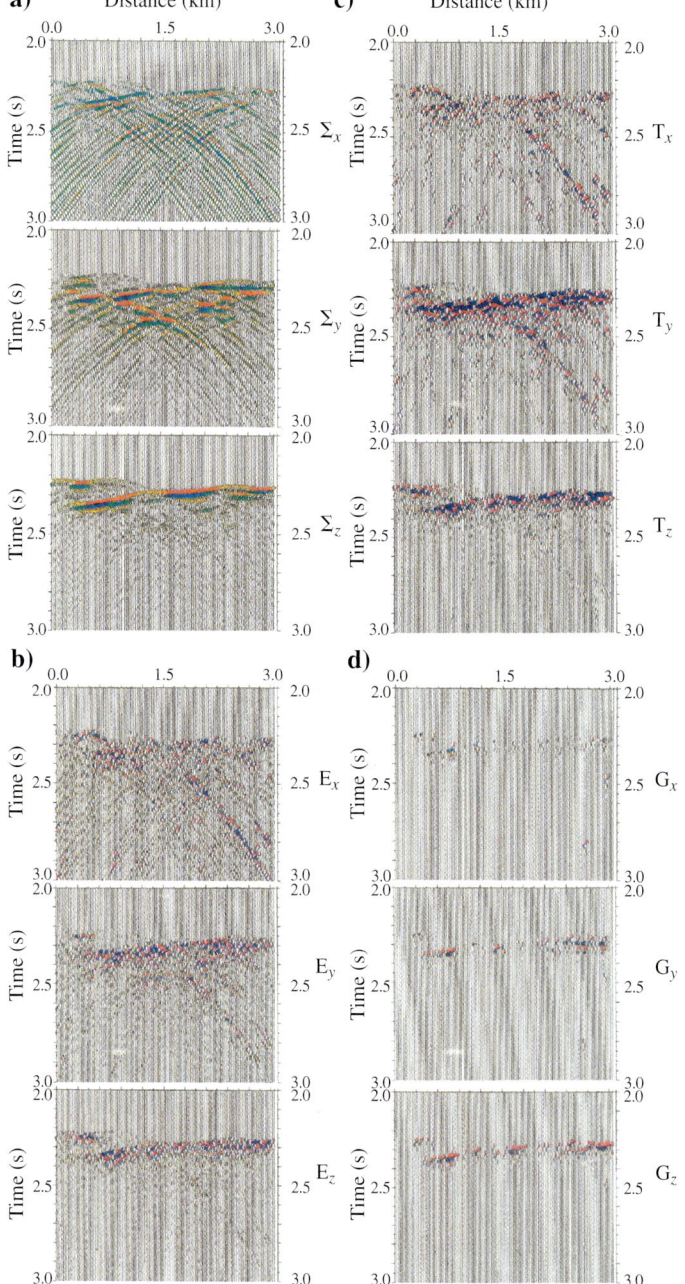

Figure 39. Normalized sections of (a) total wavefield (Σ), (b) tip waves (T), (c) edge waves (E), and (d) reflected waves (G) along line 1. The Berlage signal is 20 Hz (from Aizenberg et al., 1992b).

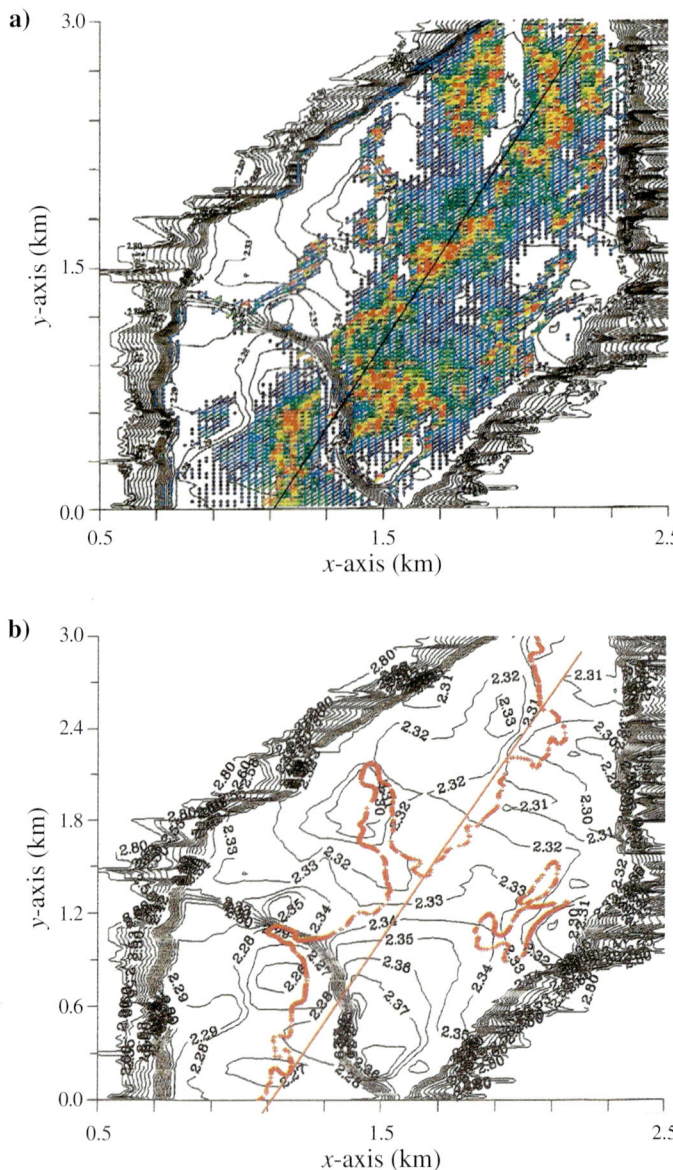

Figure 40. (a) Distribution of points representing tip waves (small dots), edge waves (lines, triangle side) and reflections (large dots), respectively, for 125 shots. Amplitudes are color-coded according to a seismic scale. Threshold levels are at 25% of amplitude for individual waves using a 50-Hz Berlage signal. (b) Footpoints of normal rays (NIP) obtained by two-point ray tracing (Hanyga, 1991) are shown for comparison. Notice that the NIP reflection points correlate with the clusters of high amplitudes (from Aizenberg et al., 1992b).

the line reflect changes in local curvature, with high-amplitude signals from the concave part of the surface and low amplitude from the convex part. Compared with the noisy appearance of individual wave-type sections (Figure 39), the stacked results Σ are much more coherent.

In Figure 40, we have stacked the DRPs of 125 shots along line 1. The superposition method offers a complete distribution of amplitude (and arrival time) of signals backscattered from the target. Isolated clusters of points coincide with fault planes, which are illuminated. White spots correspond to areas in the shadow. Tip waves dominate the distribution. However, we notice some differences in their distribution. In particular, there is a high-amplitude DRP at several locations where no NIP points have been traced. Because rays are the fundamental components of the TWS approach, similarities in the pattern of geometric waves in the DRPs and the distribution of ray footpoints might not be surprising. Almost all of the ray footpoints can be recognized by their equivalents in the DRP, although the opposite is not true.

Although the ray-tracing algorithm (Hanyga, 1991; Hanyga and Pajchel, 1995) searches for a neighborhood solution of an initial ray or previous rays in the case of successful hits, TWS involves computation of the response from each grid cell of the interface. Thus, TWS provides a more complete solution, although it is an order of magnitude higher in computational cost than ray tracing is.

Line 2

Line 2 is located at the southeast edge of the horst block. The seismic response along this line (Figure 41) has a less obvious interpretation than the previous line. Strong diffractions from the edges of fault A and the horst edge interfere with more coherent response because of the horst top.

Line 3

The synthetic section obtained for line 3 (Figure 42), starting from west along the main fault A, is dominated by the close proximity to the faults A, D, and E (Figure 38). In the vertical component seismogram Σ_z, reflections from upthrown and downthrown sides of the faults are reproduced well. The fault planes of A, D, and E give minor contributions to the seismograms. Reflection from the sides and diffraction from the near and distant sides and from distant faults are significant. The latter is more evident from the stacked DRP.

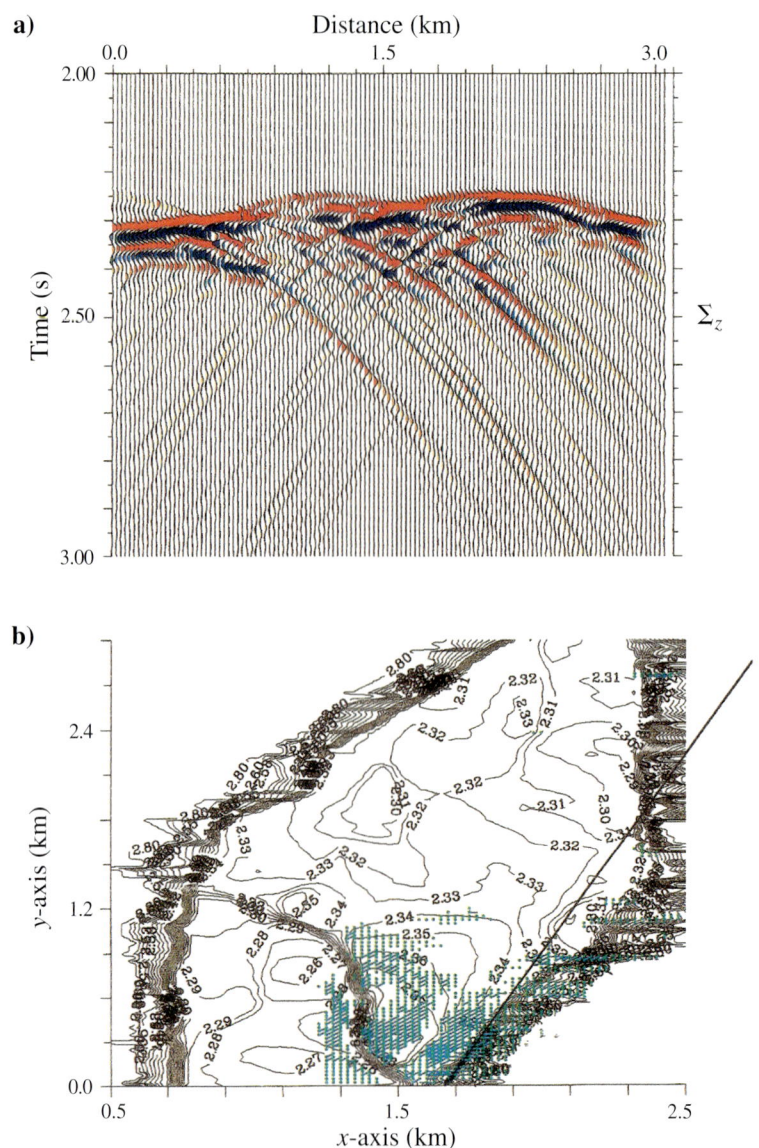

Figure 41. (a) Total P-wave field Σ_z normalized section for line 2 and (b) distribution of individual wave contributions from the first 10 shots of the seismic line (from Aizenberg et al., 1992b).

Survey design and interpretation of seismic response by the superposition method

As shown in the previous section, the recorded seismic response from boundaries with strong topographic features can be a combination of

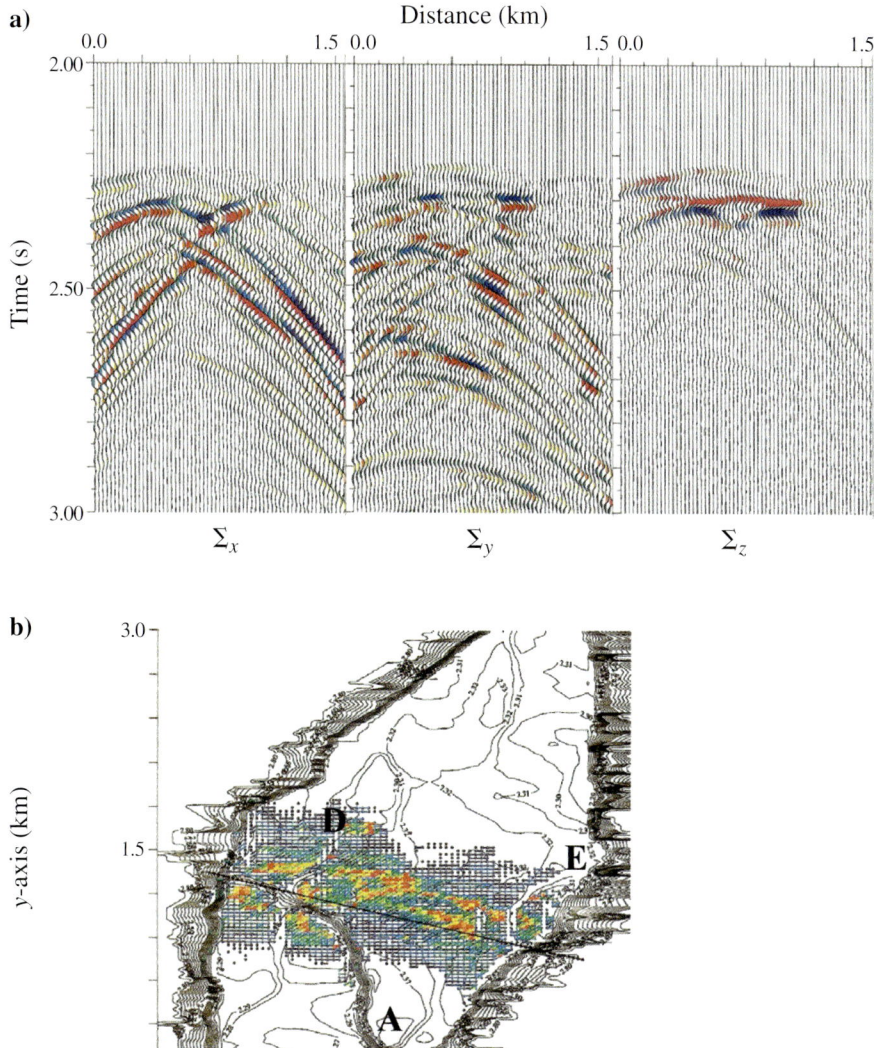

Figure 42. (a) Total P-wavefield. Normalized sections for line 3 of displacement components Σ_x (inline), Σ_y (crossline), and Σ_z (vertical). The Berlage signal is 20 Hz. (b) Corresponding DRPs and the locations of faults A, D, and E (see detail in Figure 38) (from Aizenberg et al., 1992b).

near-vertical reflections and events originating from far-offset locations. Thus, the resulting seismic section can reveal complex patterns that are difficult to interpret without the ability to identify their origin in the model. TWS modeling offers an opportunity to solve this interpretation problem because a one-to-one correspondence exists between the location of the DRP and its signature in the synthetic section. In the following, we present examples from modeling work in the North Sea, where the TWS method has been applied in survey design and interpretation of data.

We consider the model presented in Figure 43, where three seismic lines (a) are taken perpendicular to a major thrust fault, with the downthrown surface to the west-northwest and the hanging wall to the east southeast. From the distribution of DRPs, it is obvious that a 2-km line separation is far too large to allow for sufficiently detailed imaging of the reflector. A line separation of 0.5 km would be more appropriate for mapping based on a 2D seismic survey, although the chosen shooting direction seems appropriate for imaging the fault. On the other hand, a line direction coinciding with the fault has obvious disadvantages, as shown in this model (Figure 43b), where equally strong responses originate from either side of the line. These responses arrive with short spacing in the recorded seismogram (Figure 44) and thus cannot be used to interpret structural features in a conventional migration approach.

The power of TWS is demonstrated clearly in Figure 44, where individual events in the seismic section can be identified by their origin on the reflector and vice versa. The main events from the downthrown side (labeled 1a through 1d and 3 through 5) and the energy arriving from the fault (labeled 6 through 10) are all significant seismic events that might be difficult to identify by using conventional imaging techniques. Moreover, we can generate realistic synthetic seismograms where edge diffractions and caustics are handled correctly in 3D models at low computational cost compared with that of 3D finite-difference modeling. TWS, when extended to multilayer models, is a potentially powerful tool to combine survey design and processing sequence. In particular, optimal line spacing, shooting direction, and migration procedure can enhance the details in the complexity of a given model.

Summary and recent developments

As demonstrated by the above examples, the TWS method has several attractive features:

1) Geometric aspects of the ray method, a valuable feature for interpretation and survey planning, are preserved.
2) Computed seismograms for realistic models look realistic but are still interpretable because of number 1 above.

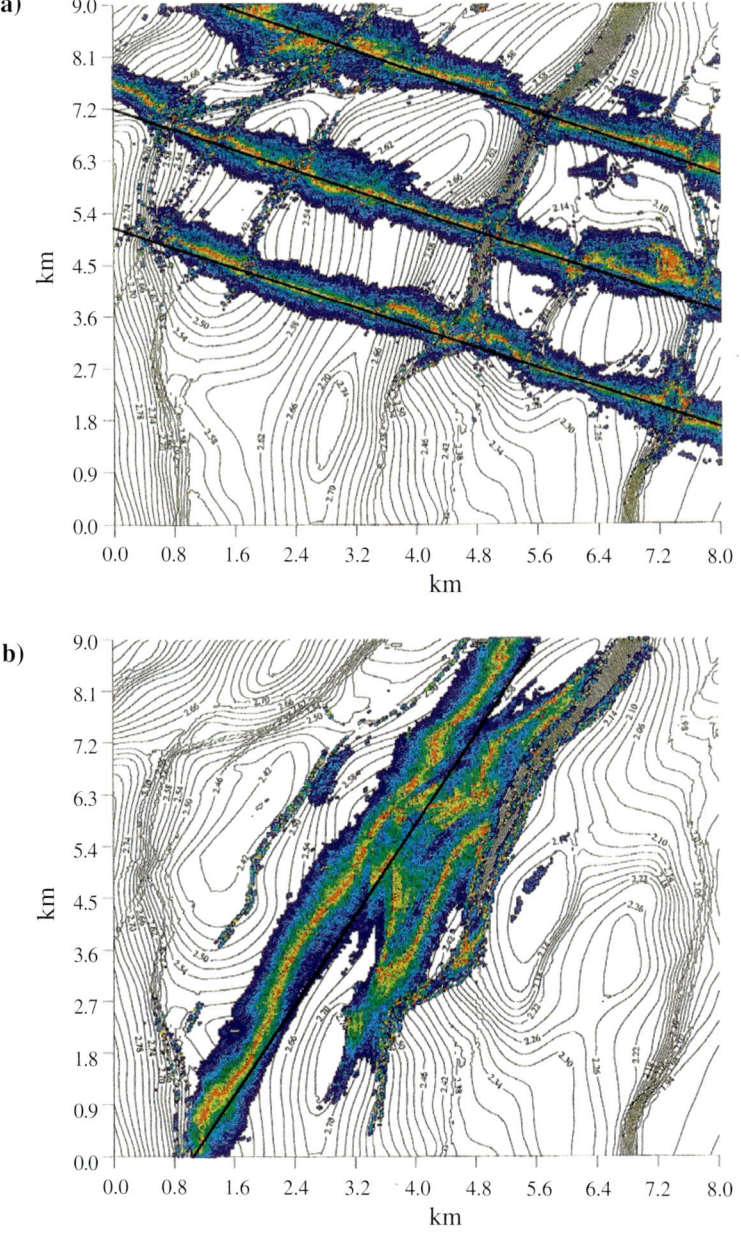

Figure 43. (a) Structural map with DRPs for three zero-offset seismic lines perpendicular to the main structural features and (b) for one seismic line along the fault on the downthrown side.

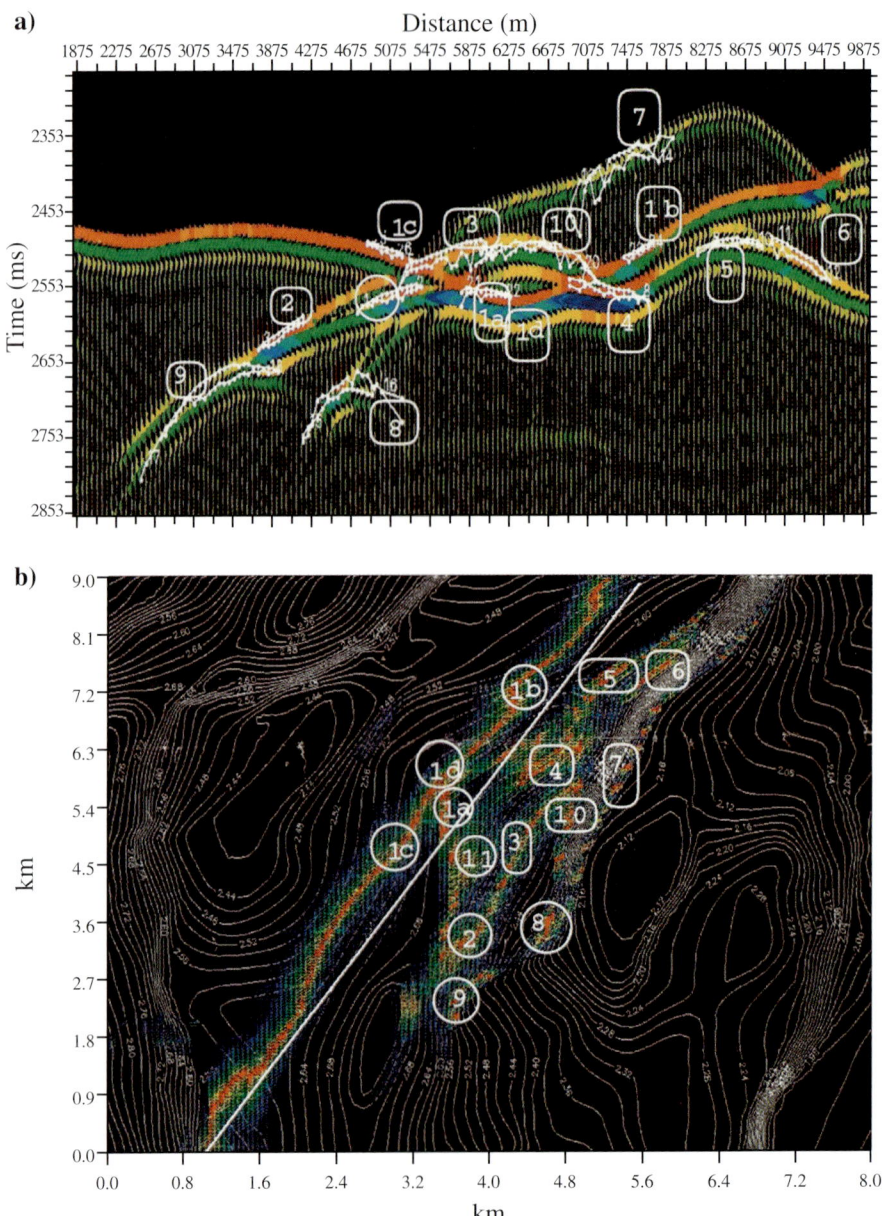

Figure 44. Details of DRP from Figure 43 showing (a) the seismic section and (b) off-line response patterns. One-to-one mapping between DRPs and seismic events is one of the useful features of the tip-wave superposition approach. Individual events are identified and labeled in both domains.

3) Significantly less computational effort is required than with pure numerical solution of the wave equation, e.g., finite difference. This implies that large-scale 3D simulation still can be done in a workstation environment.

4) Because each grid of an interface can be treated independently, computations can be shared conveniently between processors in a multiprocessor (SIMD or MIMD) environment.

Klem-Musatov et al. (2004) and Klem-Musatov et al. (2005a, 2005b) obtain a more rigorous theoretical basis to improve the TWS method for the scalar case. Aizenberg and Ayzenberg (2002) and Aizenberg et al. (2005) develop the same for the acoustic case. Preliminary results for the elastic case are represented in Ayzenberg et al. (2008b). Essentially, the improvements consist of introducing the effective (integrated) reflection/transmission coefficients instead of the plane-wave coefficients to eliminate artifacts (diffractions) caused by discontinuities of the plane-wave coefficient at critical angles. Accurate modeling of head waves is thus feasible.

As a result, extension of the theoretical basis of TWS to multilayer models was developed as the multiple tip-wave superposition (MTWS) method for an acoustic case by Ayzenberg et al. (2007b, 2008a). Aizenberg et al. (1996), Aizenberg et al. (2004), Aizenberg et al. (2006), Ayzenberg et al. (2007a, 2007b, 2008a), and Ayzenberg et al. (2008b) present the results of using the improved TWS and MTWS methods in seismic modeling. The advantage of using the improved TWS method in AVO analysis for long offsets is presented in Skopintseva et al. (2007) and Skopintseva et al. (2008).

Ayzenberg et al. (2008b) present extension to transverse isotropic (TI) media. As refinements of the theory and numerical algorithms evolve at the current rate and are demonstrated further by practical examples, we hope that the MTWS method will gain popularity.

Conclusion

Seismic diffractions are inevitable parts of the recorded wavefield scattered from complex structural settings. Thus, they carry back to the surface information that can be exploited to enhance the resolution of details underground. Diffraction phenomena can appear in different forms. In this book, we have considered only those that are caused by the edges and vertices of seismic interfaces — edge and tip waves. We limited ourselves by the approximated description of these waves only in the neighborhoods of the shadow boundaries of regular reflections and transmissions (boundary-layer approximation).

The essence of this approximation can be explained as follows: The kinematics of waves under consideration is determined by the well-known Keller's laws of edge and vertex diffraction. Their amplitudes are obtained by means of multiplication of amplitudes of regular reflections/transmissions (or their analytic continuations in shadow zones) by the attenuation functions of edge and tip diffractions, which depend only on phase shifts between fronts of waves under consideration at the observation point. This book is devoted to derivation of these functions.

It is natural to look for the attenuation function of edge diffraction by means of direct integration of the equation of motion. In the case of the scalar wave equation, the problem can be reduced easily to integration of Kummer's differential equation, which has two known linearly independent solutions. One of them leads to the known description of a wave in the framework of asymptotic ray theory, although one of the simplest linear combination of both solutions leads to the edge wave. However, one can look for the attenuation function of edge diffraction, considering the problem of smoothing discontinuity of the reflected (transmitted) wave at its shadow boundary.

This approach can be reduced to the simplest problem of singular integral equations (the Sohotsky-Plemelj problem). In this case, the edge wave plays the part of a correction that provides continuity of the total wavefield at the shadow boundary. In the boundary-layer approximation, both methods of finding the attenuation function give the same result. However, the latter approach is simpler and more efficient because it does not require the integration of equations of motion. Therefore, one can find the attenuation function of tip diffraction, considering the problem of smoothing discontinuity of the edge wave at its secondary shadow boundaries caused by the breakpoint of the diffracting edge. The simplicity of the boundary-layer approximation allows us to use it for solving direct and inverse seismic problems such as localization of diffracting objects (for example, small-throw faults, pinch-outs, etc.) and estimation of their parameters.

Edge and tip waves can be used as a basis to describe a more general diffraction phenomena. Use of these waves allows us to approximate curvilinear interfaces by piecewise plane surfaces for mathematical modeling of reflections and transmissions by the method of singular surface integrals (Kirchhoff's integral) in caustic zones. In this case, diffracted waves are regularization corrections that, with a sufficiently small approximation grid, minimize error in the scattered field caused by replacement of the curvilinear interface with a piecewise surface. This is the underlying concept of the tip-wave superposition method and its extensions. We hope this book can be useful for further development of methods to solve direct and inverse seismic problems.

References

Abramowitz, M., and I. A. Stegun, 1972, Handbook of mathematical functions: U. S. Government Printing Office.

Achenbach, J. D., 1973, Wave propagation in elastic solids: North Holland Publishing Co., 425.

Aizenberg A. M., 1982, Scattering of seismic waves by broken edge of a flat boundary: Soviet Geology and Geophysics, **23**, 74–82.

———, 1992, A self-similar conformal analog of the wave equation in 3D non-homogeneous space: Russian Geology and Geophysics, **33**, 116–121.

———, 1993a, Special function of eddy diffusion equation in 3D inhomogeneous space: Russian Geology and Geophysics, **34**, 107–114.

———, 1993b, A system of irregular fundamental solutions to the wave equation in a 3D inhomogeneous medium: Russian Geology and Geophysics, **34**, 105–113.

Aizenberg, A. M., and M. A. Ayzenberg, 2002, Simulation of interference wavefield trace reflected/transmitted by plane interface of two acoustic media (in Russian): Dynamics of solid media: Proceedings of the Russian Academy of Sciences, **121**, 50–55.

Aizenberg, A. M., M. A. Ayzenberg, H. B. Helle, K. D. Klem-Musatov, and J. Pajchel, 2004, Modeling of single reflection by tip wave superposition method using effective coefficient: 66th Annual Conference and Exhibition, EAGE, Extended Abstracts, http://www.earthdoc.org/detail.php?paperid=P187&edition=3, accessed June 16, 2008.

Aizenberg, A. M., M. A. Ayzenberg, H. B. Helle, and J. Pajchel, 2005, Reflection and transmission of acoustic wavefields at a curved interface of two inhomogeneous media (in Russian): Continuum Dynamics, Acoustics of Inhomogeneous Media, **123**, 73–79.

Aizenberg, A. M., H. B. Helle, K. D. Klem-Musatov, and J. Pajchel, 1992a, Seismic simulation by the tip wave superposition method in complex geological models: SEG/Moscow 1992 International Conference and Exposition, Technical Abstracts, 236–237.

Aizenberg, A. M., H. B. Helle, K. D. Klem-Musatov, and J. Pajchel, 1996, The tip wave superposition method based on the refraction transform: 58th Annual Conference and Exhibition, EAGE, Extended Abstracts, C001.

Aizenberg, A. M., and K. D. Klem-Musatov, 1980, Calculation of wavefields by the method of superposition of the edge waves: Soviet Geology and Geophysics, **21**, 79–94.

Aizenberg, A. M., K. D. Klem-Musatov, M. A. Ayzenberg, H. B. Helle, and J. Pajchel, 2006, Integral reflection-transmission operators instead of the reflection-transmission coefficient — A possible perspective to increase the resolving capacity of seismic exploration: Russian Geology and Geophysics, **47**, 537–546.

Aizenberg, A.M., K. D. Klem-Musatov, J. Pajchel, and H. B. Helle, 1992b, Seismic simulation by the tip wave superposition method in a complex 3D model: Open report, Norsk Hydro Research Centre.

Ayzenberg, M. A., A. M. Aizenberg, H. B. Helle, K. D. Klem-Musatov, J. Pajchel, and B. Ursin, 2007a, 3D diffraction modeling of singly scattered acoustic wavefields based on the combination of surface integral propagators and transmission operators: Geophysics, **72**, no. 5, SM19–SM34.

——, 2007b, 3D Acoustic Green's function modeling in multilayered overburden: 69th Annual Conference and Exhibition, EAGE, Extended Abstracts, P293.

Ayzenberg, M. A., A. M. Aizenberg, H. B. Helle, K. D. Klem-Musatov, J. Pajchel, and B. Ursin, 2008a, 3D modeling of acoustic Green's function in layered media with diffracting edges: 70th Annual EAGE Conference and Exhibition, Extended Abstracts, P052.

Ayzenberg, M. A., I. Tsvankin, A. M. Aizenberg, and B. Ursin, 2008b, Effective reflection coefficients for curved interfaces in TI media: 70th Annual EAGE Conference and Exhibition, Extended Abstracts, P346.

Bleistein, N., and R. A. Handelsman, 1987, Asymptotic expansions of integrals: Dover Publications.

Born, M., and E. Wolf, 2000, Principles of optics: Cambridge University Press.

Červený, V., 2001, Seismic ray theory: Cambridge University Press.

Červený, V., I. A. Molotkov, and I. Pšenčík, 1977, Ray method in seismology: Univerzita Karlova.

Červený, V., and R. Ravindra, 1971, Theory of seismic head waves: Toronto University Press.

Courant, R., and D. Hilbert, 1966, Methods of mathematical physics, 2 v.: Interscience.

Dettman, J. W., 1965, Applied complex variables: Macmillan.

Druzhinin, A. B., 1988, Kinematic law of diffraction at the edge in an anisotropic medium: Soviet Geology and Geophysics, **29**, 108–111.

——, 1990, Edge waves in an anisotropic medium: Soviet Geology and Geophysics, **31**, 118–130.

——, 1991, Diffraction of seismic waves by an irregular edge of the interface of arbitrary elastic media: Soviet Geology and Geophysics, **32**, 97–105.

Druzhinin, A. B., and A. M. Aizenberg, 1990, Asymptotic solutions of equations of motion of an anisotropic medium: Soviet Geology and Geophysics, **31**, 121–132.

Felsen, L. B., and N. Marcuvitz, 1973, Radiation and scattering of waves: Prentice Hall, 888.

Fock, V. A., 1965, Electromagnetic diffraction and propagation problems: Pergamon Press, 414.

Frøyland, L. A., H. B. Helle, P. Riste, and I. Sandø, 1988, A study of detectability and resolution of a complex structure by seismic survey modelling and synthetic data processing: Norwegian Petroleum Society (NPF) Biannual Geophysical Seminar.

Gallop, J. B., and F. Hron, 1998, Diffractions and boundary conditions in asymptotic ray theory: Geophysical Journal International, **133**, 413–418.

Goldin, S. V., 2005, Introduction to geometrical seismics (in Russian): Novosibirsk State University.

Hanyga A., l99l, Ray tracing in the case of multiple-valued traveltimes: 6lst Annual International Meeting, SEG, Expanded Abstracts, 1517–1521.

Hanyga, A., and J. Pajchel, 1995, Point-to-curve ray tracing in a real reservoir model: Geophysical Prospecting, **43**, 859–872.

Hoffmann, H. J., D. Klaeschen, W. Rabbel, and E. R. Flueh, 1993, Ray tracing based migration operators for complex media and discontinuities by the edge-wave method: 55th Annual Conference and Exhibition, EAGE, Extended Abstracts, B038.

Hron, F., and G. H. Chan, 1995, Tutorial on the numerical modelling of edge-diffracted waves by the ray method: Studia Geophysics et Geodetica, **39**, 103–137.

Hönl, H., A. W. Maue, and K. Westphal, 1961, Theorie der beugung: Springer-Verlag.

Keller, J. B., 1962, A geometrical theory of diffraction: Journal of the Optical Society of America, **52**, 116–130.

Khaidukov, V., E. Landa, and T. J. Moser, 2004, Diffraction imaging by focusing-defocusing: An outlook on seismic superresolution: Geophysics, **69**, 1478–1490.

Klaeschen, D., W. Rabbel, and E. R. Flueh, 1994, An automated ray method for diffraction modelling in complex media: Geophysical Journal International, **116**, 23–38.

Klem-Musatov, K. D., 1980, Theory of edge waves and its applications in seismology (in Russian): Nauka.

———, 1981, Asymptotic formulae for the amplitude of waves scattered by a salient point on a diffracting edge: Soviet Geology and Geophysics, **22**, no. 3, 108–114.

———, 1994, Theory of seismic diffractions: SEG Open File Publications No. 1.

Klem-Musatov, K. D., and A. M. Aizenberg, 1984, Ray method and the theory of edge waves: Geophysical Journal of the Royal Astronomical Society, **79**, 35–50.

———, 1985, Seismic modelling by methods of the theory edge waves: Journal of Geophysics, **57**, 90–105.

———, 1989, The edge wave superposition method (2D scalar problem): Geophysical Journal International, **99**, 351–367.

Klem-Musatov, K. D, A. M. Aizenberg, H. B. Helle, and J. Pajchel, 2004, Reflection and transmission at curvilinear interface in terms of surface integrals: Wave Motion, **39**, 77–92.

———, 2005a, Reflection and transmission in multilayered media in terms of surface integrals: Wave Motion, **41**, 293–305.

Klem-Musatov, K. D., A. M. Aizenberg, H. B. Helle, J. Pajchel, and M.A. Ayzenberg, 2005b, Reflection and transmission at curvilinear interface in terms of surface integrals: Workshop, Seismic Waves in Laterally Inhomogeneous Media VI, http://rebel.ig.cas.cz/activities/Present2005/km_aa.ppt, accessed June 16, 2008.

Klem-Musatov, K. D., A. M. Aizenberg, and G. A. Klem-Musatova, 1982, An algorithm for mathematical modeling of three-dimensional diffraction fields: Soviet Geology and Geophysics, **23**, 116–121.

Klem-Musatov, K. D., G. L. Kovalevsky, and V. G. Chernyakov, 1976, Seismic anomalies associated with local disturbances: Soviet Geology and Geophysics, **17**, 103–121.

Klem-Musatov, K. D., G. L. Kovalevsky, V. G. Chernyakov, and L. G. Maksimov, 1975, Mathematical modelling of diffraction of seismic waves in angular regions: Soviet Geology and Geophysics, **16**, 88–97.

Klem-Musatov, K. D., G. L. Kovalevsky, and L. R. Tokmulina, 1972, On the intensity of waves diffracted on the edge: Geologiya i Geofizika, **5**, 82–92.

Landa, E., and S. Keydar, 1998, Seismic monitoring of diffraction images for detection of local heterogeneities: Geophysics, **63**, 1093–1100.

Landa, E. I., and L. A. Maksimov, 1980, Testing an algorithm for the separation of low-amplitude faults: Soviet Geology and Geophysics, **21**, 108–113.

Landa, E., V. Shtivelman, and B. Gelchinsky, 1987, A method for detection of diffracted waves on common-offset sections: Geophysical Prospecting, **35**, 359–373.

Luneva, M. N., 1996, Application of the edge wave superposition method: Pure and Applied Geophysics, **148**, 113–136.

Luneva, M. N., and S. M. Kharlamov, 1990, Parallel physical and mathematical modelling of wave fields passing through a complex interface of media: Soviet Geology and Geophysics, **31**, 101–106.

Muskhelishvili, N. I., 1953, Singular integral equations: Noordhoff.

Pajchel, J., H. B. Helle, and L. A. Frøyland, 1987, A 2D seismic modelling software package: User's manual: Open report, Norsk Hydro Research Centre.

———, 1988, Computation of seismic edge diffractions by the theory of edge waves: Open report, Norsk Hydro Research Centre, Seismic Waves in Inhomogeneous Media III Workshop.

———, 1989, The theory of edge waves applied to the modelling of seismic diffractions in VSP: 51st Annual Conference and Exhibition, EAEG, Extended Abstracts, C39.

Skopintseva, L. V., T. V. Nefedkina, M. A. Ayzenberg, and A. M. Aizenberg, 2007, An approach to the AVO-inversion problem based on the effective reflection coefficients: 69th Annual Conference and Exhibition, EAGE, Extended Abstracts, P354.

Skopintseva L. V., M. A. Ayzenberg, M. Landrø, T. V. Nefedkina, and A. M. Aizenberg, 2008, AVO inversion of long-offset synthetic PP data based on effective reflection coefficients. 70th Annual Conference and Exhibition, EAGE, Extended Abstracts, P347.

Wang, X., and D. Waltham, 1995, The stable-beam seismic modelling method: Geophysical Prospecting, **43**, 939–961.

Index